George William Jones

Five-place Logarithms

George William Jones

Five-place Logarithms

ISBN/EAN: 9783337131128

Printed in Europe, USA, Canada, Australia, Japan

Cover: Foto ©ninafisch / pixelio.de

More available books at **www.hansebooks.com**

FIVE-PLACE LOGARITHMS

BY

GEORGE WILLIAM JONES, A.M.,

PROFESSOR OF MATHEMATICS IN CORNELL UNIVERSITY.

These Tables are specially designed for use in the class-room; they are arranged in the usual way, and it is assumed that the teacher will give all needed explanations. To promote the detection of errors, a dollar will be paid for the first notice of each error. Address Professor Jones at Ithaca, N. Y.

FIRST EDITION.

ITHACA, N. Y.
GEORGE W. JONES.
1896.

100	0	1	2	3	4	5	6	7	8	9	Differences.
100	00 000	043	087	130	173	217	260	303	346	389	
01	432	475	518	561	604	647	689	732	775	817	
02	860	903	945	988	*030	*072	*115	*157	*199	*242	
03	01 284	326	368	410	452	494	536	578	620	662	
04	703	745	787	829	870	912	953	995	*036	*078	
05	02 119	160	202	243	284	325	366	407	449	490	
06	531	572	612	653	694	735	776	816	857	898	
07	938	979	*019	*060	*100	*141	*181	*222	*262	*302	
08	03 342	383	423	463	503	543	583	623	663	703	
09	743	782	822	862	902	941	981	*021	*060	*100	
110	04 139	179	218	258	297	336	376	415	454	493	
11	532	571	610	650	689	727	766	805	844	883	
12	922	961	999	*038	*077	*115	*154	*192	*231	*269	
13	05 308	346	385	423	461	500	538	576	614	652	
14	690	729	767	805	843	881	918	956	994	*032	
15	06 070	108	145	183	221	258	296	333	371	408	
16	446	483	521	558	595	633	670	707	744	781	
17	819	856	893	930	967	*004	*041	*078	*115	*151	
18	07 188	225	262	298	335	372	408	445	482	518	
19	555	591	628	664	700	737	773	809	846	882	
120	07 918	954	990	*027	*063	*099	*135	*171	*207	*243	
21	08 279	314	350	386	422	458	493	529	565	600	
22	636	672	707	743	778	814	849	884	920	955	
23	991	*026	*061	*096	*132	*167	*202	*237	*272	*307	
24	09 342	377	412	447	482	517	552	587	621	656	
25	691	726	760	795	830	864	899	934	968	*003	
26	10 037	072	106	140	175	209	243	278	312	346	
27	380	415	449	483	517	551	585	619	653	687	
28	721	755	789	823	857	890	924	958	992	*025	
29	11 059	093	126	160	193	227	261	294	327	361	
130	11 394	428	461	494	528	561	594	628	661	694	
31	727	760	793	826	860	893	926	959	992	*024	
32	12 057	090	123	156	189	222	254	287	320	352	
33	385	418	450	483	516	548	581	613	646	678	
34	710	743	775	808	840	872	905	937	969	*001	
35	13 033	066	098	130	162	194	226	258	290	322	
36	354	386	418	450	481	513	545	577	609	640	
37	672	704	735	767	799	830	862	893	925	956	
38	988	*019	*051	*082	*114	*145	*176	*208	*239	*270	
39	14 301	333	364	395	426	457	489	520	551	582	
140	14 613	644	675	706	737	768	799	829	860	891	
41	922	953	983	*014	*045	*076	*106	*137	*168	*198	
42	15 229	259	290	320	351	381	412	442	473	503	
43	534	564	594	625	655	685	715	746	776	806	
44	836	866	897	927	957	987	*017	*047	*077	*107	
45	16 137	167	197	227	256	286	316	346	376	406	
46	435	465	495	524	554	584	613	643	673	702	
47	732	761	791	820	850	879	909	938	967	997	
48	17 026	056	085	114	143	173	202	231	260	289	
49	319	348	377	406	435	464	493	522	551	580	
150	0	1	2	3	4	5	6	7	8	9	Differences.

Differences (proportional parts):

	44	43	42	41	40
1	4	4	4	4	4
2	9	9	8	8	8
3	13	13	13	12	12
4	18	17	17	16	16
5	22	22	21	21	20
6	26	26	25	25	24
7	31	30	29	29	28
8	35	34	34	33	32
9	40	39	38	37	36

	39	38	37	36	35
1	4	4	4	4	4
2	8	8	7	7	7
3	12	11	11	11	11
4	16	15	15	14	14
5	20	19	19	18	18
6	23	23	22	22	21
7	27	27	26	25	25
8	31	30	30	29	28
9	35	34	33	32	32

	34	33	32	31	30
1	3	3	3	3	3
2	7	7	6	6	6
3	10	10	10	9	9
4	14	13	13	12	12
5	17	17	16	16	15
6	20	20	19	19	18
7	24	23	22	22	21
8	27	26	26	25	24
9	31	30	29	28	27

	29	28	27	26	25
1	3	3	3	3	3
2	6	6	5	5	5
3	9	8	8	8	8
4	12	11	11	10	10
5	15	14	14	13	13
6	17	17	16	16	15
7	20	20	19	19	18
8	23	22	22	21	20
9	26	25	24	23	23

	24	23	22	21
1	2	2	2	2
2	5	5	4	4
3	7	7	7	6
4	10	9	9	8
5	12	12	11	11
6	14	14	13	13
7	17	16	15	15
8	19	18	18	17
9	22	21	20	19

150	0	1	2	3	4	5	6	7	8	9
150	17 609	638	667	696	725	754	782	811	840	869
51	898	926	955	984	*013	*041	*070	*099	*127	*156
52	18 184	213	241	270	298	327	355	384	412	441
53	469	498	526	554	583	611	639	667	696	724
54	752	780	808	837	865	893	921	949	977	*005
55	19 033	061	089	117	145	173	201	229	257	285
56	312	340	368	396	424	451	479	507	535	562
57	590	618	645	673	700	728	756	783	811	838
58	866	893	921	948	976	*003	*030	*058	*085	*112
59	20 140	167	194	222	249	276	303	330	358	385
160	20 412	439	466	493	520	548	575	602	629	656
61	683	710	737	763	790	817	844	871	898	925
62	952	978	*005	*032	*059	*085	*112	*139	*165	*192
63	21 219	245	272	299	325	352	378	405	431	458
64	484	511	537	564	590	617	643	669	696	722
65	748	775	801	827	854	880	906	932	958	985
66	22 011	037	063	089	115	141	167	194	220	246
67	272	298	324	350	376	401	427	453	479	505
68	531	557	583	608	634	660	686	712	737	763
69	789	814	840	866	891	917	943	968	994	*019
170	23 045	070	096	121	147	172	198	223	249	274
71	300	325	350	376	401	426	452	477	502	528
72	553	578	603	629	654	679	704	729	754	779
73	805	830	855	880	905	930	955	980	*005	*030
74	24 055	080	105	130	155	180	204	229	254	279
75	304	329	353	378	403	428	452	477	502	527
76	551	576	601	625	650	674	699	724	748	773
77	797	822	846	871	895	920	944	969	993	*018
78	25 042	066	091	115	139	164	188	212	237	261
79	285	310	334	358	382	406	431	455	479	503
180	25 527	551	575	600	624	648	672	696	720	744
81	768	792	816	840	864	888	912	935	959	983
82	26 007	031	055	079	102	126	150	174	198	221
83	245	269	293	316	340	364	387	411	435	458
84	482	505	529	553	576	600	623	647	670	694
85	717	741	764	788	811	834	858	881	905	928
86	951	975	998	*021	*045	*068	*091	*114	*138	*161
87	27 184	207	231	254	277	300	323	346	370	393
88	416	439	462	485	508	531	554	577	600	623
89	646	669	692	715	738	761	784	807	830	852
190	27 875	898	921	944	967	989	*012	*035	*058	*081
91	28 103	126	149	171	194	217	240	262	285	307
92	330	353	375	398	421	443	466	488	511	533
93	556	578	601	623	646	668	691	713	735	758
94	780	803	825	847	870	892	914	937	959	981
95	29 003	026	048	070	092	115	137	159	181	203
96	226	248	270	292	314	336	358	380	403	425
97	447	469	491	513	535	557	579	601	623	645
98	667	688	710	732	754	776	798	820	842	863
99	885	907	929	951	973	994	*016	*038	*060	*081
200	0	1	2	3	4	5	6	7	8	9

NATURAL LOGARITHMS.

Num.	Log.	Num.	Log.
0	∞	50	3.91 202
1	0.00 000	51	3.93 183
2	0.69 315	52	3.95 124
3	1.09 861	53	3.97 029
4	1.38 629	54	3.98 898
5	1.60 944	55	4.00 733
6	1.79 176	56	4.02 585
7	1.94 591	57	4.04 305
8	2.07 944	58	4.06 044
9	2.19 722	59	4.07 754
10	2.30 259	60	4.09 484
11	2.39 790	61	4.11 087
12	2.48 491	62	4.12 713
13	2.56 495	63	4.14 313
14	2.63 906	64	4.15 888
15	2.70 805	65	4.17 439
16	2.77 259	66	4.18 965
17	2.83 321	67	4.20 469
18	2.89 037	68	4.21 951
19	2.94 444	69	4.23 411
20	2.99 573	70	4.24 850
21	3.04 452	71	4.26 268
22	3.09 104	72	4.27 667
23	3.13 549	73	4.29 046
24	3.17 805	74	4.30 407
25	3.21 888	75	4.31 749
26	3.25 810	76	4.33 073
27	3.29 584	77	4.34 381
28	3.33 220	78	4.35 671
29	3.36 730	79	4.36 945
30	3.40 120	80	4.38 203
31	3.43 399	81	4.39 445
32	3.46 574	82	4.40 672
33	3.49 651	83	4.41 884
34	3.52 636	84	4.43 082
35	3.55 535	85	4.44 265
36	3.58 352	86	4.45 435
37	3.61 092	87	4.46 591
38	3.63 759	88	4.47 734
39	3.66 356	89	4.48 864
40	3.68 888	90	4.49 981
41	3.71 357	91	4.51 086
42	3.73 767	92	4.52 179
43	3.76 120	93	4.53 260
44	3.78 419	94	4.54 329
45	3.80 666	95	4.55 388
46	3.82 864	96	4.56 435
47	3.85 015	97	4.57 471
48	3.87 120	98	4.58 497
49	3.89 182	99	4.59 512
50	3.91 202	100	4.60 517

NATURAL LOGARITHMS.

200	0	1	2	3	4	5	6	7	8	9	Differences.	
200	30 103	125	146	168	190	211	233	255	276	298		
01	320	341	363	384	406	428	449	471	492	514		
02	535	557	578	600	621	643	664	685	707	728		22 21
03	750	771	792	814	835	856	878	899	920	942	1	2 2
04	963	984	*006	*027	*048	*069	*091	*112	*133	*154	2	4 4
											3	7 6
05	31 175	197	218	239	260	281	302	323	345	366	4	9 8
06	387	408	429	450	471	492	513	534	555	576	5	11 11
07	597	618	639	660	681	702	723	744	765	785	6	13 13
08	806	827	848	869	890	911	931	952	973	994	7	15 15
09	32 015	035	056	077	098	118	139	160	181	201	8	18 17
											9	20 19
210	32 222	243	263	284	305	325	346	'366	387	408		
11	428	449	469	490	510	531	552	572	593	613		
12	634	654	675	695	715	736	756	777	797	818		20 19
13	838	858	879	899	919	940	960	980	*001	*021	1	2 2
14	33 041	062	082	102	122	143	163	183	203	224	2	4 4
											3	6 6
15	244	264	284	304	325	345	365	385	405	425	4	8 8
16	445	465	486	506	526	546	566	586	606	626	5	10 10
17	646	666	686	706	726	746	766	786	806	826	6	12 11
18	846	866	885	905	925	945	965	985	*005	*025	7	14 13
19	34 044	064	084	104	124	143	163	183	203	223	8	16 15
											9	18 17
220	34 242	262	282	301	321	341	361	380	400	420		
21	439	459	479	498	518	537	557	577	596	616		18 17
22	635	655	674	694	713	733	753	772	792	811	1	2 2
23	830	850	869	889	908	928	947	967	986	*005	2	4 3
24	35 025	044	064	083	102	122	141	160	180	199	3	5 5
											4	7 7
25	218	238	257	276	295	315	334	353	372	392	5	9 9
26	411	430	449	468	488	507	526	545	564	583	6	11 10
27	603	622	641	660	679	698	717	736	755	774	7	13 12
28	793	813	832	851	870	889	908	927	946	965	8	14 14
29	984	*003	*021	*040	*059	*078	*097	*116	*135	*154	9	16 15
230	36 173	192	211	229	248	267	286	305	324	342		
31	361	380	399	418	436	455	474	493	511	530		16 15
32	549	568	586	605	624	642	661	680	698	717	1	2 2
33	736	754	773	791	810	829	847	866	884	903	2	3 3
34	922	940	959	977	996	*014	*033	*051	*070	*088	3	5 5
											4	6 6
35	37 107	125	144	162	181	199	218	236	254	273	5	8 8
36	291	310	328	346	365	383	401	420	438	457	6	10 9
37	475	493	511	530	548	566	585	603	621	639	7	11 11
38	658	676	694	712	731	749	767	785	803	822	8	13 12
39	840	858	876	894	912	931	949	967	985	*003	9	14 14
240	38 021	039	057	075	093	112	130	148	166	184		14
41	202	220	238	256	274	292	310	328	346	364	1	1
42	382	399	417	435	453	471	489	507	525	543	2	3
43	561	578	596	614	632	650	668	686	703	721	3	4
44	739	757	775	792	810	828	846	863	881	899	4	6
											5	7
45	917	934	952	970	987	*005	*023	*041	*058	*076	6	8
46	39 094	111	129	146	164	182	199	217	235	252	7	10
47	270	287	305	322	340	358	375	393	410	428	8	11
48	445	463	480	498	515	533	550	568	585	602	9	13
49	620	637	655	672	690	707	724	742	759	777		
250	0	1	2	3	4	5	6	7	8	9	Differences.	

250	0	1	2	3	4	5	6	7	8	9
250	39 704	811	829	846	863	881	898	915	933	950
51	967	985	*002	*019	*037	*054	*071	*088	*106	*123
52	40 140	157	175	192	209	226	243	261	278	295
53	312	329	346	364	381	398	415	432	449	466
54	483	500	518	535	552	569	586	603	620	637
55	654	671	688	705	722	739	756	773	790	807
56	824	841	858	875	892	909	926	943	960	976
57	993	*010	*027	*044	*061	*078	*095	*111	*128	*145
58	41 162	179	196	212	229	246	263	280	296	313
59	330	347	363	380	397	414	430	447	464	481
260	41 497	514	531	547	564	581	597	614	631	647
61	664	681	697	714	731	747	764	780	797	814
62	830	847	863	880	896	913	929	946	963	979
63	996	*012	*029	*045	*062	*078	*095	*111	*127	*144
64	42 160	.177	193	210	226	243	259	275	292	308
65	325	341	357	374	390	406	423	439	455	472
66	488	504	521	537	553	570	586	602	619	635
67	651	667	684	700	716	732	749	765	781	797
68	813	830	846	862	878	894	911	927	943	959
69	975	991	*008	*024	*040	*056	*072	*088	*104	*120
270	43 136	152	169	185	201	217	233	249	265	281
71	297	313	329	345	361	377	393	409	425	441
72	457	473	489	505	521	537	553	569	584	600
73	616	632	648	664	680	696	712	727	743	759
74	775	791	807	823	838	854	870	886	902	917
75	933	949	965	981	996	*012	*028	*044	*059	*075
76	44 091	107	122	138	154	170	185	201	217	232
77	248	264	279	295	311	326	342	358	373	389
78	404	420	436	451	467	483	498	514	529	545
79	560	576	592	607	623	638	654	669	685	700
280	44 716	731	747	762	778	793	809	824	840	855
81	871	886	902	917	932	948	963	979	994	*010
82	45 025	040	056	071	086	102	117	133	148	163
83	179	194	209	225	240	255	271	286	301	317
84	332	347	362	378	393	408	423	439	454	469
85	484	500	515	530	545	561	576	591	606	621
86	637	652	667	682	697	712	728	743	758	773
87	788	803	818	834	849	864	879	894	909	924
88	939	954	969	984	*000	*015	*030	*045	*060	*075
89	46 090	105	120	135	150	165	180	195	210	225
290	46 240	255	270	285	300	315	330	345	359	374
91	389	404	419	434	449	464	479	494	509	523
92	538	553	568	583	598	613	627	642	657	672
93	687	702	716	731	746	761	776	790	805	820
94	835	850	864	879	894	909	923	938	953	967
95	982	997	*012	*026	*041	*056	*070	*085	*100	*114
96	47 129	144	159	173	188	202	217	232	246	261
97	276	290	305	319	334	349	363	378	392	407
98	422	436	451	465	480	494	509	524	538	553
99	567	582	596	611	625	640	654	669	683	698
300	0	1	2	3	4	5	6	7	8	9

TEN-PLACE LOGARITHMS.

NUM.	LOG.
2	.301 029 9957
3	.477 121 2547
5	.698 970 0043
7	.845 098 0400
11	.041 392 6852
13	.113 943 3523
17	.230 448 9214
19	.278 753 6010
23	.361 727 8360
29	.462 397 9979
31	.491 361 6938
37	.568 201 7241
41	.612 783 8567
43	.633 468 4556
47	.672 097 8579
53	.724 275 8696
59	.770 852 0116
61	.785 329 8350
67	.826 074 8027
71	.851 258 3437
73	.863 322 8601
79	.897 627 0918
83	.919 078 0924
89	.949 390 0066
97	.986 771 7843
101	.004 321 3788
103	.012 837 2247
107	.029 383 7777
109	.037 426 4979
113	.053 078 4435
127	.103 803 7210
131	.117 271 2957
137	.136 720 5672
139	.143 014 8008
149	.173 186 2684
151	.178 976 9473
157	.195 899 6524
163	.212 187 6044
167	.222 716 4711
173	.238 046 1031
179	.252 853 0310
181	.257 678 5749
191	.281 033 3672
193	.285 557 3090
197	.294 466 2262
199	.298 853 0764
211	.324 282 4553
223	.348 304 8630
227	.356 025 8572
229	.359 835 4823
233	.367 855 9210

TEN-PLACE LOGARITHMS.

300	0	1	2	3	4	5	6	7	8	9
300	47 712	727	741	756	770	784	799	813	828	842
01	857	871	885	900	914	929	943	958	972	986
02	48 001	015	029	044	058	073	087	101.	116	130
03	144	159	173	187	202	216	230	244	259	273
04	287	302	316	330	344	359	373	387	401	416
05	430	444	458	473	487	501	515	530	544	558
06	572	586	601	615	629	643	657	671	686	700
07	714	728	742	756	770	785	799	813	827	841
08	855	869	883	897	911	926	940	954	968	982
09	996	*010	*024	*038	*052	*066	*080	*094	*108	*122
310	49 136	150	164	178	192	206	220	234	248	262
11	276	290	304	318	332	346	360	374	388	402
12	415	429	443	457	471	485	499	513	527	541
13	554	568	582	596	610	624	638	651	665	679
14	693	707	721	734	748	762	776	790	803	817
15	831	845	859	872	886	900	914	927	941	955
16	969	982	996	*010	*024	*037	*051	*065	*079	*092
17	50 106	120	133	147	161	174	188	202	215	229
18	243	256	270	284	297	311	325	338	352	365
19	379	393	406	420	433	447	461	474	488	501
320	50 515	529	542	556	569	583	596	610	623	637
21	651	664	678	691	705	718	732	745	759	772
22	786	799	813	826	840	853	866	880	893	907
23	920	934	947	961	974	987	*001	*014	*028	*041
24	51 055	068	081	095	108	121	135	148	162	175
25	188	202	215	228	242	255	268	282	295	308
26	322	335	348	362	375	388	402	415	428	441
27	455	468	481	495	508	521	534	548	561	574
28	587	601	614	627	640	654	667	680	693	706
29	720	733	746	759	772	786	799	812	825	838
330	51 851	865	878	891	904	917	930	943	957	970
31	983	996	*009	*022	*035	*048	*061	*075	*088	*101
32	52 114	127	140	153	166	179	192	205	218	231
33	244	257	270	284	297	310	323	336	349	362
34	375	388	401	414	427	440	453	466	479	492
35	504	517	530	543	556	569	582	595	608	621
36	634	647	660	673	686	699	711	724	737	750
37	763	776	789	802	815	827	840	853	866	879
38	892	905	917	930	943	956	969	982	994	*007
39	53 020	033	046	058	071	084	097	110	122	135
340	53 148	161	173	186	199	212	224	237	250	263
41	275	288	301	314	326	339	352	364	377	390
42	403	415	428	441	453	466	479	491	504	517
43	529	542	555	567	580	593	605	618	631	643
44	656	668	681	694	706	719	732	744	757	769
45	782	794	807	820	832	845	857	870	882	895
46	908	920	933	945	958	970	983	995	*008	*020
47	54 033	045	058	070	083	095	108	120	133	145
48	156	170	183	195	208	220	233	245	258	270
49	283	295	307	320	332	345	357	370	382	394
350	0	1	2	3	4	5	6	7	8	9

Differences.

15	14	13	12	11	10
1 2	1 1	1 1	1 1	1 1	1 1
2 3	2 3	2 3	2 2	2 2	2 2
3 5	3 4	3 4	3 4	3 3	3 3
4 6	4 6	4 5	4 5	4 4	4 4
5 8	5 7	5 7	5 6	5 6	5 5
6 9	6 8	6 8	6 7	6 7	6 6
7 11	7 10	7 9	7 8	7 9	7 7
8 12	8 11	8 10	8 10	8 0	8 8
9 14	9 13	9 12	9 11	9 10	9 9

350	0	1	2	3	4	5	6	7	8	9	WEIGHTS AND MEASURES.
350	54 407	419	432	444	456	469	481	494	506	518	**LENGTH.**
51	531	543	555	568	580	593	●●5	617	630	642	Num. Log.
52	654	667	679	691	704	716	728	741	753	765	Inches in 1 meter.
53	777	790	802	814	827	839	851	864	876	888	39.3700 1.59 517
54	900	913	925	937	949	962	974	986	998	*011	Meters in 1 inch.
											.02 540 8.40 483
55	55 023	035	047	060	072	084	096	108	121	133	Feet in 1 meter.
											3/4, 3.2 809 0.51 598
56	145	157	169	182	194	206	218	230	242	255	Meters in 1 foot.
57	267	279	291	303	315	328	340	352	364	376	4/13, .30 480 9.48 402
58	388	400	413	425	437	449	461	473	485	497	Yards in 1 meter.
											13/12, 1.0 936 0.03 886
59	509	522	534	546	558	570	582	594	606	618	Meters in 1 yard.
											12/13, .91 440 9.96 114
360	55 630	642	654	666	678	691	703	715	727	739	Miles in 1 kilometer.
61	751	763	775	787	799	811	823	835	847	859	5/8, .62 137 9.79 885
62	871	883	895	907	919	931	943	955	967	979	Kilometers in 1 mile.
											8/5, 1.6 093 0.20 665
63	.991	*003	*015	*027	*038	*050	*062	*074	*086	*098	**AREA.**
64	56 110	122	134	146	158	170	182	194	205	217	Sq. inches in 1 sq. meter.
											1550.0 3.19 083
65	229	241	253	265	277	289	301	312	324	336	Sq. meters in 1 sq. inch.
											.00 065 6.80 967
66	348	360	372	384	396	407	419	431	443	455	Sq. feet in 1 sq. meter.
67	467	478	490	502	514	526	538	549	561	573	10.764 1.03 197
											Sq. meters in 1 sq. foot.
68	585	597	608	620	632	644	656	667	679	691	1/11, .09 290 8.96 803
69	703	714	726	738	750	761	773	785	797	808	Sq. yards in 1 sq. meter.
											6/5, 1.1 960 0.07 773
370	56 820	832	844	855	867	879	891	902	914	926	Sq. meters in 1 sq. yard.
71	937	949	961	972	984	996	*008	*019	*031	*043	5/6, .83 613 9.92 227
72	57 054	066	078	089	101	113	124	136	148	159	**VOLUME.**
											Cu. inches in 1 cu. cm.
73	171	183	194	206	217	229	241	252	264	276	.06 102 8.78 550
74	287	299	310	322	334	345	357	368	380	392	Cu. cm. in 1 cu. inch.
											16.387 1.21 450
75	403	415	426	438	449	461	473	484	496	507	Cu. feet in 1 cu. meter.
76	519	530	542	553	565	576	588	600	611	623	35.314 1.54 795
											Cu. meters in 1 cu. foot.
77	634	646	657	669	680	692	703	715	726	738	.02 632 8.45 205
78	749	761	772	784	795	807	818	830	841	852	Cu. yards in 1 cu. meter.
											4/3, 1.3 079 0.11 659
79	864	875	887	898	910	921	933	944	955	967	Cu. meters in 1 cu. yard.
											3/4, .76 456 9.88 341
380	57 978	990	*001	*013	*024	*035	*047	*058	*070	*081	**CAPACITY.**
81	58 092	104	115	127	138	149	161	172	184	195	Fluid ounce in 1 cu. cm.
											.03 381 8.52 909
82	206	218	229	240	252	263	274	286	297	309	Cu. cm. in 1 fluid ounce.
83	320	331	343	354	365	377	388	399	410	422	29.574 1.47 091
											U. S. gallons in 1 liter.
84	433	444	456	467	478	490	501	512	524	535	.26 417 9.42 189
											Liters in 1 U. S. gallon.
85	546	557	569	580	591	602	614	625	636	647	3.7 854 0.57 812
86	659	670	681	692	704	715	726	737	749	760	U. S. bushels in 1 hectoliter.
87	771	782	794	805	816	827	838	850	861	872	2.8 377 0.45 297
88	883	894	906	917	928	939	950	961	973	984	Hectoliters in 1 U. S. bushel.
											.35 239 9.54 708
89	995	*006	*017	*028	*040	*051	*062	*073	*084	*096	**WEIGHTS.**
											Grains in 1 gram.
390	59 106	118	129	140	151	162	173	184	195	207	15.482 1.18 643
91	218	229	240	251	262	273	284	295	306	318	Grams in 1 grain.
											.06 480 5.81 157
92	329	340	351	362	373	384	395	406	417	428	Oz. troy in 1 gram.
93	439	450	461	472	483	494	506	517	528	539	.03 215 5.50 719
											Grams in 1 oz. troy.
94	550	561	572	583	594	605	616	627	638	649	31.103 1.49 281
											Oz. avoirdupois in 1 gram.
95	660	671	682	693	704	715	726	737	748	759	.03 527 6.54 745
96	770	780	791	802	813	824	835	846	857	868	Grains in 1 oz. avoirdupois.
											28.350 1.45 255
97	879	890	901	912	923	934	945	956	966	977	Pounds av. in 1 kilogram.
98	988	999	*010	*021	*032	*043	*054	*065	*076	*086	11/5, 2.2 046 0.34 383
											Kilograms in 1 pound av.
99	60 097	108	119	130	141	152	163	173	184	195	5/11, .45 359 9.65 667
400	**0**	**1**	**2**	**3**	**4**	**5**	**6**	**7**	**8**	**9**	**WEIGHTS AND MEASURES.**

400	0	1	2	3	4	5	6	7	8	9
400	60 206	217	228	239	249	260	271	282	293	304
01	314	325	336	347	358	369	379	390	401	412
02	423	433	444	455	466	477	487	498	509	520
03	531	541	552	563	574	584	595	606	617	627
04	638	649	660	670	681	692	703	713	724	735
05	. 746	756	767	778	788	799	810	821	831	842
06	853	863	874	885	895	906	917	927	938	949
07	959	970	981	991	*002	*013	*023	*034	*045	*055
08	61 066	077	087	098	109	119	130	140	151	162
09	172	183	194	204	215	225	236	247	257	268
410	61 278	289	300	310	321	331	342	352	363	374
11	384	395	405	416	426	437	448	458	469·	479
12	490	500	511	521	532	542	553	563	574	584
13	595	606	616	627	637	648	658	669	679	690
14	700	711	721	731	742	752	763	773	784	794
15	805	815	826	836	847	857	868	878	888	899
16	909	920	930	941	951	962	972	982	993	*003
17	62 014	024	034	045	055	066	076	086	097	107
18	118	128	138	149	159	170	180	190	201	211
19	221	232	242	252	263	273	284	294	304	315
420	62 325	335	346	356	366	377	387	397	408	418
21	428	439	449	459	469	480	490	500	511	521
22	531	542	552	562	572	583	593	603	613	624
23	634	644	655	665	675	685	696	706	716	726
24	737	747	757	767	778	788	798	808	818	829
25	839	849	859	870	880	890	900	910	921	931
26	941	951	961	972	982	992	*002	*012	*022	*033
27	63 043	053	063	073	083	094	104	114	124	134
28	144	155	165	175	185	195	205	215	225	236
29	246	256	266	276	286	296	306	317	327	337
430	63 347	357	367	377	387	397	407	417	428	438
31	448	458	468	478	488	498	508	518	528	538
32	548	558	568	579	589	599	609	619	629	639
33	649	659	669	679	689	699	709	719	729	739
34	749	759	769	779	789	799	809	819	829	839
35	849	859	869	879	889	899	909	919	929	939
36	949	959	969	979	988	998	*008	*018	*028	*038
37	64 048	058	068	078	·088	098	108	118	128	137
38	147	157	167	177	187	197	207	217	227	237
39	246	256	266	276	286	296	306	316	326	335
440	64 345	355	365	375	385	395	404	414	424	434
41	444	454	464	473	483	493	503	513	523	532
42	542	552	562	572	582	591	601	611	621	631
43	640	650	660	670	680	689	699	709	719	729
44	738	748	758	768	777	787	797	807	816	826
45	836	846	856	865	875	885	895	904	914	924
46	933	943	953	963	972	982	992	*002	*011	*021
47	65 031	040	050	060	070	079	089	099	108	118
48	128	137	147	157	167	176	186	196	205	215
49	225	234	244	254	263	273	283	292	302	312
450	0	1	2	3	4	5	6	7	8	9

Differences.

11
1 1
2 2
8 8
4 4
5 6
6 7
7 8
8 9
9 10

10
1 1
2 2
3 3
4 4
5 5
6 6
7 7
8 8
9 9

9
1 1
2 2
8 8
4 4
5 5
6 6
7 6
8 7
9 8

8
1 1
2 2
8 2
4 3
5 4
6 5
7 6
8 6
9 7

450	0	1	2	3	4	5	6	7	8	9
450	65 321	331	341	350	360	369	379	389	398	408
51	418	427	437	447	456	466	475	485	495	504
52	514	523	533	543	552	562	571	581	591	600
53	610	619	629	639	648	658	667	677	686	696
54	706	715	725	734	744	753	763	772	782	792
55	801	811	820	830	839	849	858	868	877	887
56	896	906	916	925	935	944	954	963	973	982
57	992	*001	*011	*020	*030	*039	*049	*058	*068	*077
58	66 087	096	106	115	124	134	143	153	162	172
59	181	191	200	210	219	229	238	247	257	266
460	66 276	285	295	304	314	323	332	342	351	361
61	370	380	389	398	408	417	427	436	446	455
62	464	474	483	492	502	511	521	530	539	549
63	558	567	577	586	596	605	614	624	633	642
64	652	661	671	680	689	699	708	717	727	736
65	745	755	764	773	783	792	801	811	820	829
66	839	848	857	867	876	885	894	904	913	922
67	932	941	950	960	969	978	987	997	*006	*015
68	67 025	034	043	052	062	071	080	089	099	108
69	117	127	136	145	154	164	173	182	191	201
470	67 210	219	228	237	247	256	265	274	284	293
71	302	311	321	330	339	348	357	367	376	385
72	394	403	413	422	431	440	449	459	468	477
73	486	495	504	514	523	532	541	550	560	569
74	578	587	596	605	614	624	633	642	651	660
75	669	679	688	697	706	715	724	733	742	752
76	761	770	779	788	797	806	815	825	834	843
77	852	861	870	879	888	897	906	916	925	934
78	943	952	961	970	979	988	997	*006	*015	*024
79	68 034	043	052	061	070	079	088	097	106	115
480	68 124	133	142	151	160	169	178	187	196	205
81	215	224	233	242	251	260	269	278	287	296
82	305	314	323	332	341	350	359	368	377	386
83	395	404	413	422	431	440	449	458	467	476
84	485	494	502	511	520	529	538	547	556	565
85	574	583	592	601	610	619	628	637	646	655
86	664	673	681	690	699	708	717	726	735	744
87	753	762	771	780	789	797	806	815	824	833
88	842	851	860	869	878	886	895	904	913	922
89	931	940	949	958	966	975	984	993	*002	*011
490	69 020	028	037	046	055	064	073	082	090	099
91	108	117	126	135	144	152	161	170	179	188
92	197	205	214	223	232	241	249	258	267	276
93	285	294	302	311	320	329	338	346	355	364
94	373	381	390	399	408	417	425	434	443	452
95	461	469	478	487	496	504	513	522	531	539
96	548	557	566	574	583	592	601	609	618	627
97	636	644	653	662	671	679	688	697	705	714
98	723	732	740	749	758	767	775	784	793	801
99	810	819	827	836	845	854	862	871	880	888
500	0	1	2	3	4	5	6	7	8	9

MATHEMATICAL CONSTANTS.

LOGARITHMS.

$e = 1 + 1 + 1/2! + 1/3! + 1/4! +$
= · · · 2.71828 18284 59045
$M = \log_{10} e =$ · · .43429 44819
$1/M = \log_e 10 =$ · 2.30256 50930

PI.

$\pi =$ · · 3.14159 26535 89793
$1/\pi =$ · · 0.31830 98861 83791
$\sqrt{\pi} =$ · · 1.77245 38509 05516
$1/\sqrt{\pi} =$ · 0.56418 95835 47756
$\pi^2 =$ · · 9.86960 44010 89358
$1/\pi^2 =$ · · 0.10132 11836 42338
$\pi^3 =$ · · 31.00627 66802 99820
$1/\pi^3 =$ · · 0.03225 15344 33199
$\pi^4 =$ · · 97.40909 10340 09437
$1/\pi^4 =$ · · 0.01026 59922 54664
$\log_e \pi =$ · · · 1.14472 98858
$\log_{10} \pi =$ · · · 0.49714 98727
log arc 1° = · 8.24187 78076
log arc 1' = · 6.46372 61172
log arc 1" = · 4.68557 48665

RADIUS MEASURE.

$R° = 180°/\pi =$ 57°.29577 95181
$R' = 180·60/\pi$ · · · ·
= · · 3437".74677 07849
$R'' = 180·60·60/\pi =$ · ·
= · · 206264".80624 70964
$\log R° =$ · 1.75812 26324
$\log R' =$ · · 3.53627 38828
$\log R'' =$ · · 5.31442 51832

TRIGONOMETRIC RATIOS.

sin 1° = · .01745 24064 37284
sin 1' = · .00029 08882 04568
sin 1" = · .00000 48481 36811
log sin 1° = · · 8.24185 53184
log sin 1' = · · · 6.46372 61111
log sin 1" = · · · 4.68557 48666

PWRS OF 1", 1', 1°, IN RADIANS.

$1'' =$ .0⁵48481 36611 09386
$(1'')^2 =$ .0¹⁰23504 43054
$(1'')^2/2! =$ .0¹⁰1175221527
$(1'')^3 =$ .0¹⁵ 11396
$(1'')^3/3! =$ .0¹⁵ 1899
$1' =$ .0³29 08882 08665 72160
$(1')^2 =$ .0⁷846 15049 94075
$(1')^2/2! =$ .0⁷428 07974 97088
$(1')^3 =$ .0¹⁰24618 79210
$(1')^3/3! =$ .0¹¹4102 29702
$(1')^4 =$ .0¹⁴71 5066
$(1')^4/4! =$ .0¹⁵ 29838
$(1')^5 =$ .0¹⁷7208
$(1')^5/5! =$ .0¹⁰2
$1° =$ .01745 32925 19948 29577
$(1°)^2 =$ .0³30 46174 19786 70600
$(1°)^2/2! =$ .0³15 23087 09808 35430
$(1°)^3 =$ .0⁵53105 76084 20779
$(1°)^3/3! =$ .0⁵8860 96155 70130
$(1°)^4 =$ .0⁷927 91772 48751 16477
$(1°)^4/4! =$ .0⁸88 66328 85156 29987
$(1°)^5 =$ .0⁸16 19521 94779 59000
$(1°)^5/5! =$ .0¹⁰18496 01623 16826
$(1°)^6 =$ .0¹⁰28265 99029 73508
$(1°)^6/6! =$ .0¹²89 25581 98574
$(1°)^7 =$ .0¹²493 83459 70255
$(1°)^7/7! =$ .0¹⁵9785 85486
$(1°)^8 =$ .0¹⁴8 61081 80820 94968
$(1°)^8/8! =$ .0¹²21 35484 80859
$(1°)^9 =$ .0¹⁵15027 83120 37484
$(1°)^{10} =$ .0¹⁷202 28518 29393
$(1°)^{11} =$ .0¹⁹4 51773 91668

MATHEMATICAL CONSTANTS.

500	0	1	2	3	4	5	6	7	8	9
500	69 897	906	914	923	932	940	949	958	966	975
01	984	992	*001	*010	*018	*027	*036	*044	*053	*062
02	70 070	079	088	096	105	114	122	131	140	148
03	157	165	174	183	191	200	209	217	226	234
04	243	252	260	269	278	286	295	303	312	321
05	329	338	346	355	364	372	381	389	398	406
06	415	424	432	441	449	458	467	475	484	492
07	501	509	518	526	535	544	552	561	569	578
08	586	595	603	612	621	629	638	646	655	663
09	672	680	689	697	706	714	723	731	740	749
510	70 757	766	774	783	791	800	808	817	825	834
11	842	851	859	868	876	885	893	902	910	919
12	927	935	944	952	961	969	978	986	995	*003
13	71 012	020	029	037	046	054	063	071	079	088
14	096	105	113	122	130	139	147	155	164	172
15	181	189	198	206	214	223	231	240	248	257
16	265	273	282	290	299	307	315	324	332	341
17	349	357	366	374	383	391	399	408	416	425
18	433	441	450	458	466	475	483	492	500	508
19	517	525	533	542	550	559	567	575	584	592
520	71 600	609	617	625	634	642	650	659	667	675
21	684	692	700	709	717	725	734	742	750	759
22	767	775	784	792	800	809	817	825	834	842
23	850	858	867	875	883	892	900	908	917	925
24	933	941	950	958	966	975	983	991	999	*008
25	72 016	024	032	041	049	057	066	074	082	090
26	099	107	115	123	132	140	148	156	165	173
27	181	189	198	206	214	222	230	239	247	255
28	263	272	280	288	296	304	313	321	329	337
29	346	354	362	370	378	387	395	403	411	419
530	72 428	436	444	452	460	469	477	485	493	501
31	509	518	526	534	542	550	558	567	575	583
32	591	599	607	616	624	632	640	648	656	665
33	673	681	689	697	705	713	722	730	738	746
34	754	762	770	779	787	795	803	811	819	827
35	835	843	852	860	868	876	884	892	900	908
36	916	925	933	941	949	957	965	973	981	989
37	997	*006	*014	*022	*030	*038	*046	*054	*062	*070
38	73 078	086	094	102	111	119	127	135	143	151
39	159	167	175	183	191	199	207	215	223	231
540	73 239	247	255	263	272	280	288	296	304	312
41	320	328	336	344	352	360	368	376	384	392
42	400	408	416	424	432	440	448	456	464	472
43	480	488	496	504	512	520	528	536	544	552
44	560	568	576	584	592	600	608	616	624	632
45	640	648	656	664	672	679	687	695	703	711
46	719	727	735	743	751	759	767	775	783	791
47	799	807	815	823	830	838	846	854	862	870
48	878	886	894	902	910	918	926	933	941	949
49	957	965	973	981	989	997	*005	*013	*020	*028
550	0	1	2	3	4	5	6	7	8	9

Differences.

	9
1	1
2	2
3	3
4	4
5	5
6	5
7	6
8	7
9	8

	8
1	1
2	2
3	2
4	3
5	4
6	5
7	6
8	6
9	7

	7
1	1
2	1
3	2
4	3
5	4
6	4
7	5
8	6
9	6

550	0	1	2	3	4	5	6	7	8	9
550	74 036	044	052	060	068	076	084	092	099	107
51	115	123	131	139	147	155	162	170	178	.186
52	194	202	210	218	225	233	241	249	257	265
53	273	280	288	296	304	312	320	327	335	343
54	351	359	367	374	382	390	398	406	414	421
55	429	437	445	453	461	468	476	484	492	500
56	507.	515	523	531	539	547	554	562	570	578
57	586	593	601	609	617	624	632	640	648	656
58	663	671	679	687	695	702	710	718	726	733
59	741	749	757	764	772	780	788	796	803	811
560	74 819	827	834	842	850	858	865	873	881	889
61	896	904	912	920	927	935	943	950	958	966
62	974	981	989	997	*005	*012	*020	*028	*035	*043
63	75 051	059	066	074	082	089	097	105	113	120
64	128	136	143	151	159	166	174	182	189	197
65	205	213	220	228	236	243	251	259	266	274
66	282	289	297	305	312	320	328	335	343	351
67	358	366	374	381	389	397	404	412	420	427
68	435	442	450	458	465	473	481	488	496	504
69	511	519	526	534	542	549	557	565	572	580
570	75 587	595	603	610	618	626	633	641	648	656
71	664	671	679	686	694	702	709	717	724	732
72	740	747	755	762	770	778	785	793	800	808
73	815	823	831	838	846	853	861	868	876	884
74	891	899	906	914	921	929	937	944	952	959
75	967	974	982	989	997	*005	*012	*020	*027	*035
76	76 042	050	057	065	072	080	087	095	103	110
77	118	125	133	140	148	155	163	170	178	185
78	193	200	208	215	223	230	238	245	253	260
79	268	275	283	290	298	305	313	320	328	335
580	76 343	350	358	365	373	380	388	395	403	410
81	418	425	433	440	448	455	462	470	477	485
82	492	500	507	515	522	530	537	545	552	559
83	567	574	582	589	597	604	612	619	626	634
84	641	649	656	664	671	678	686	693	701	708
85	716	723	730	738	745	753	760	768	775	782
86	790	797	805	812	819	827	834	842	849	856
87	864	871	879	886	893	901	908	916	923	930
88	938	945	953	960	967	975	982	989	997	*004
89	77 012	019	026	034	041	048	056	063	070	078
590	77 085	093	100	107	115	122	129	137	144	151
91	159	166	173	181	188	195	203	210	217	225
92	232	240	247	254	262	269	276	283	291	298
93	305	313	320	327	335	342·	349	357	364	371
94	379	386	393	401	408	415	422	430	437	444
95	452	459	466	474	481	488	495	503	510	517
96·	77 525	532	539	546	554	561	568	576	583	590
97	597	605	612	619	627	634	641	648	656	663
98	670	677	685	692	699	706	714	721	728	735
99	743	750	757	764	772	779	786	793	801	808
600	0	.1	2	3	4	5	6	7	8	9

MERIDIONAL PARTS.

ANG. MILES.		ANG. MILES.	
0°	0	42°	2782
1	60	43	2863
2	120	44	2946
3	180	45	3030
4	240	46	3116
5	300	47	3203
6	361	48	3292
7	421	49	3382
8	482	50	3474
9	542	51	3569
10	603	52	3665
11	664	53	3764
12	725	54	3865
13	787	55	3968
14	848	56	4074
.15	910	57	4183
16	973	58	4294
17	1085	59	4409
18	1098	60	4527
19	1161	61	4649
20	1225	62	4775
21	1289	63	4905
22	1354	64	5089
23	1419	65	5179
24	1484	66	5324
25	1550	67	5474
26	1616	68	5631
27	1684	69	5795
28	1751	70	5966
29	1819	71	6146
30	1888	72	6335
31	1958	73	6534
32	2029	74	6746
33	2100	75	6970
34	2171	76	7210
35	2244	77	7467
36	2318	78	7745
37	2398	79	8046
38	2468	80	8375
39	2545	81	8739
40	2623	82	9145
41	2702	83	9606

600	0	1	2	3	4	5	6	7	8	9
600	77 815	822	830	837	844	851	859	866	873	880
01	887	895	902	909	916	924	931	938	945	952
02	960	967	974	981	988	996	*003	*010	*017	*025
03	78 032	039	046	053	061	068	075	082	089	097
04	104	111	118	125˙	132	140	147	154	161	168
05	176	183	190	197	204	211	219	226	233	240
06	247	254	262	269	276	283	290	297	305	312
07	319	326	333	340	347	355	362	369	376	383
08	390	398	405	412	419	426	433	440	447	455
09	462	469	476	483	490	497	504	512	519	526
610	78 533	540	547	554	561	569	576	583	590	597
11	604	611	618	625	633	640	647	654	661	668
12	675	682	689	696	704	711	718	725	732	739
13	746	753	760	767	774	781	789	796	803	810
14	817	824	831	838	845	852	859	866	873	880
15	888	895	902	909	916	923	930	937	944	951
16	958	965	972	979	986	993	*000	*007	*014	*021
17	79 029	036	043	050 .	057	064	071	078	085	092
18	099	106	113	120	127	134	141	148	155	162
19	169	176	183	190	197	204	211	218	225	232
620	79 239	246	253	260	267	274	281	288	295	302
21	309	316	323	330	337	344	351	358	365	372
22	379	386	393	400	407	414	421	428	435	442
23	449	456	463	470	477	484	491	498	505	511
24	518	525	532	539	546	553	560	567	574	581
25	588	595	602	609	616	623	630	637	644	650
26	657	664	671	678	685	692	699	706	713	720
27	727	734	741	748	754	761	768	775	782	789
28	796	803	810	817	824	831	837	844	851	858
29	865	872	879	886	893	900	906	913	920	927
630	79 934	941	948	955	962	969	975	982	989	996
31	80 003	010	017	024	030	037	044	051	058	065
32	072	079	085	092	099	106	113	120	127	134
33	140	147	154	161	168	175	182	188	195	202
34	209	216	223	229	236	243	250	257	264	271
35	277	284	291	298	305	312	318	325	332	339
36	346	353	359	366	373	380	387	393	400	407
37	414	421	428	434	441	448	455	462	468	475
38	482	489	496	502	509	516	523	530	536	543
39	550	557	564	570	577	584	591	598	604	611
640	80 618	625	632	638	645	652	659	665	672	679
41	686	693	699	706	713	720	726	733	740	747
42	754	760	767	774	781	787	794	801	808	814
43	821	828	835	841	848	855	862	868	875	882
44	889	895	902	909	916	922	929	936	943	949
45	956	963	969	976	983	990	996	*003	*010	*017
46	81 023	030	037	043	050	057	064	070	077	084
47	090	097	104	111	117	124	131	137	144	151
48	158	164	171	178	184	191	198	204	211	218
49	224	231	238	245	251	258	265	271	278	285
650	0	1	2	3	4	5	6	7	8	9

Differences.

8
1	1
2	2
3	2
4	3
5	4
6	5
7	6
8	6
9	7

7
1	1
2	1
3	2
4	3
5	4
6	4
7	5
8	6
9	6

6
1	1
2	1
3	2
4	2
5	3
6	4
7	4
8	5
9	5

650	0	1	2	3	4	5	6	7	8	9
650	81 291	298	305	311	318	325	331	338	345	351
51	358	365	371	378	385	391	398	405	411	418
52	425	431	438	445	451	458	465	471	478	485
53	.491	498	505	511	518	525	531	538	544	551
54	558	564	571	578	584	591	598	604	611	617
55	624	631	637	644	651	657	664	671	677	684
56	690	697	704	710	717	723	730	737	743	750
57	757	763	770	776	783	790	796	803	809	816
58	823	829	836	842	849	856	862	869	875	882
59	889	895	902	908	915	921	928	935	941	948
660	81 954	961	968	974	981	987	994	*000	*007	*014
61	82 020	027	033	040	046	053	060	066	073	079
62	086	092	099	105	112	119	125	132	138	145
63	151	158	164	171	178	184	191	197	204	210
64	217	223	230	236	243'	249	256	263	269	276
65	282	289	295	302	308	315	321	328	334	341
66	347	354	360	367	373	380	387	393	400	406
67	413	419	426	432	439	445	452	458	465	471
68	478	484	491	497	504·	510	517	523	530	536
69	543	549	556	562	569	575	582	588	595	601
670	82 607	614	620	627	633	640	646	653	659	666
71	672	679	685	692	698	705	711	718	724	730
72	737	743	750	756	763	769	776	782	789	795
73	802	808	814	821	827	834	840	847	853	860
74	866	872	879	885	892	898	905	911	918	924
75	930	937	943	950	956	963	969	975	982	988
76	995	*001	*008	*014	*020	*027	*033	*040	*046	*052
77	83 059	065	072	078	085	091	097	104	110	117
78	123	129	136	142	149	155	161	168	174	181
79	187	193	200	206	213	219	225	232	238	245
680	83 251	257	264	270	276	283	289	296	302	308
81	315	321	327	334	340	347	353	359	366	372
82	378	385	391	398	404	410	417	423	429	436
83	442	448	455	461	467	474	480	487	493	499
84	506	512	518	525	531	537	544	550	556	563
85	569	575	582	588	594	601	607	613	620	626
86	632	639	645	651	658	664	670	677	683	689
87	696	702	708	715	721	727	734	740	746	753
88	759	765	771	778	784	790	797	803	809	816
89	822	828	835	841	847	853	860	866	872	879
690	83 885	891	897	904	910	916	923	929	935	942
91	948	954	960	967	973	979	985	992	998	*004
92	84 011	017	023	029	036	042	048	055	061	067
93	073	080	086	092	098	105	111	117	123	130
94	136	142	148	155	161	167	173	180	186	192
95	198	205	211	217	223	230	236	242	248	255
96	261	267	273	280	286	292	298	305	311	317
97	323	330	336	342	348	354	361	367	373	379
98	386	392	398	404	410	417	423	429	435	442
99	448	454	460	466	473	479	485	491	497	504
700	0	1	2	3	4	5	6	7	8	9

SQUARE ROOTS.

Num.	Root.	Num.	Root.
0	0.0000	50	7.0711
1	1.0000	51	1414
2	4142	52	2111
3	7321	53	2801
4	2.0000	54	3485
5	2361	55	4162
6	4495	56	4833
7	6458	57	5498
8	8284	58	6158
9	3.0000	59	6811
10	3.1623	60	7.7460
11	3166	61	8102
12	4641	62	8740
13	6056	63	9373
14	7417	64	8.0000
15	8730	65	0623
16	4.0000	66	1240
17	1231	67	1854
18	2426	68	2462
19	3589	69	3066
20	4.4721	70	8.3666
21	5826	71	4261
22	6904	72	4853
23	7958	73	5440
24	8990	74	6023
25	5.0000	75	6603
26	0990	76	7178
27	1962	77	7750
28	2915	78	8318
29	3852	79	8882
30	5.4772	80	8.9443
31	5678	81	9.0000
32	6569	82	0554
33	7446	83	1104
34	8310	84	1652
35	9161	85	2195
36	6.0000	86	2736
37	0828	87	3274
38	1644	88	3808
39	2450	89	4340
40	6.3246	90	9.4868
41	4031	91	5394
42	4807	92	5917
43	5574	93	6437
44	6332	94	6954
45	7082	95	7468
46	7823	96	7980
47	8557	97	8489
48	9282	98	8995
49	7.0000	99	9499
50	0711	100	10.0000

700	0	1	2	3	4	5	6	7	8	9	Differences.
700	84 510	516	522	528	535	541	547	553	559	566	
01	572	578	584	590	597	603	609	615	621	628	
02	634	640	646	652	658	665	671	677	683	689	
03	696	702	708	714	720	726	733	739	745	751	
04	757	763	770	776	782	788	794	800	807	813	
05	819	825	931	837	844	850	856	862	868	874	
06	880	887	893	899	905	911	917	924	930	936	
07	942	948	954	960	967	973	979	985	991	997	**7**
08	85 003	009	016	022	028	034	040	046	052	058	1 1 2 1
09	065	071	077	083	.089	095	101	107	114	120	3 2 4 3
710	85 126	132	138	144	150	156	163	169	175	181	5 4 6 4
11	187	193	199	205	211	217	224	230	236	242	7 5
12	248	254	260	266	272	278	285	291	297	303	8 6
13	309	315	321	327	333	339	345	352	358	364	9 6
14	370	376	382	388	394	400	406	412	418	425	
15	431	437	443	449	455	461	467	473	479	485	
16	491	497	503	509	516	522	528	534	540	546	
17	552	558	564	570	576	582	588	594	600	606	
18	612	618	625	631	637	643	649	655	661	667	
19	673	679	685	691	697	703	709	715	721	727	
720	85 733	739	745	751	757	763	769	775	781	788	
21	794	800	806	812	818	824	830	836	842	848	**6**
22	854	860	866	872	878	884	890	896	902	908	1 1
23	914	920	926	932	938	944	950	956	962	968	2 1 3 2
24	974	980	986	992	998	*004	*010	*016	*022	*028	4 2
25	86 034	040	046	052	058	064	070	076	082	088	5 3
26	094	100	106	112	118	124	130	136	141	147	6 4
27	153	159	165	171	177	183	189	195	201	207	7 4
28	213	219	225	231	237	243	249	255	261	267	8 5
29	273	279	285	291	297	303	308	314	320	326	9 5
730	86 332	338	344	350	356	362	368	374	380	386	
31	392	398	404	410	415	421	427	433	439	445	
32	451	457	463	469	475	481	487	493	499	504	
33	510	516	522	528	534	540	546	552	558	564	
34	570	576	581	587	593	599	605	611	617	623	
35	629	635	641	646	652	658	664	670	676	682	
36	688	694	700	705	711	717	723	729	735	741	**5**
37	747	753	759	764	770	776	782	788	794	800	1 1
38	806	812	817	823	829	835	841	847	853	859	2 1 3 2
39	864	870	876	882	888	894	900	906	911	917	4 2
740	86 923	929	935	941	947	953	958	964	970	976	5 3
41	982	988	994	999	*005	*011	*017	*023	*029	*035	6 3
42	87 040	046	052	058	064	070	075	081	087	093	7 4
43	099	105	111	116	122	128	134	140	146	151	8 4
44	157	163	169	175	181	186	192	198	204	210	9 5
45	216	221	227	233	239	245	251	256	262	268	
46	274	280	286	291	297	303	309	315	320	326	
47	332	338	344	349	355	361	367	373	379	384	
48	390	396	402	408	413	419	425	431	437	442	
49	448	454	460	466	471	477	483	489	495	500	
750	0	1	2	3	4	5	6	7	8	9	Differences.

750	0	1	2	3	4	5	6	7	8	9
750	87 506	512	518	523	529	535	541	547	552	558
51	564	570	576	581	587	593	599	604	610	616
52	622	628	633	639	645	651	656	662	668	674
53	679	685	691	697	703	708	714	720	726	731
54	737	743	749	754	760	766	772	777	783	789
55	795	800	806	812	818	823	829	835	841	846
56	852	858	864	869	875	881	887	892	898	904
57	910	915	921	927	933	938	944	950	955	961
58	967	973	978	984	990	996	*001	*007	*013	*018
59	88 024	030	036	041	047	053	058	064	070	076
760	88 081	087	093	098	104	110	116	121	127	133
61	138	144	150	156	161	167	173	178	184	190
62	195	201	207	213	218	224	230	235	241	247
63	252	258	264	270	275	281	287	292	298	304
64	309	315	321	326	332	338	343	349	355	360
65	366	372	377	383	389	395	400	406	412	417
66	423	429	434	440	446	451	457	463	468	474
67	480	485	491	497	502	508	513	519	525	530
68	536	542	547	553	559	564	570	576	581	587
69	593	598	604	610	615	621	627	632	638	643
770	88 649	655	660	666	672	677	683	689	694	700
71	705	711	717	722	728	734	739	745	750	756
72	762	767	773	779	784	790	795	801	807	812
73	818	824	829	835	840	846	852	857	863	868
74	874	880	885	891	897	902	908	913	919	925
75	930	936	941	947	953	958	964	969	975	981
76	986	992	997	*003	*009	*014	*020	*025	*031	*037
77	89 042	048	053	059	064	070	076	081	087	092
78	098	104	109	115	120	126	131	137	143	148
79	154	159	165	170	176	182	187	193	198	204
780	89 209	215	221	226	232	237	243	248	254	260
81	265	271	276	282	287	293	298	304	310	315
82	321	326	332	337	343	348	354	360	365	371
83	376	382	387	393	398	404	409	415	421	426
84	432	437	443	448	454	459	465	470	476	481
85	487	492	498	504	509	515	520	526	531	537
86	542	548	553	559	564	570	575	581	586	592
87	597	603	609	614	620	625	631	636	642	647
88	653	658	664	669	675	680	686	691	697	702
89	708	713	719	724	730	735	741	746	752	757
790	89 763	768	774	779	785	790	796	801	807	812
91	818	823	829	834	840	845	851	856	862	867
92	873	878	883	889	894	900	905	911	916	922
93	927	933	938	944	949	955	960	966	971	977
94	982	988	993	998	*004	*009	*015	*020	*026	*031
95	90 037	042	048	053	059	064	069	075	080	086
96	091	097	102	108	113	119	124	129	135	140
97	146	151	157	162	168	173	179	184	189	195
98	200	206	211	217	222	227	233	238	244	249
99	255	260	266	271	276	282	287	293	298	304
800	0	1	2	3	4	5	6	7	8	9

CUBE ROOTS.

Num.	Root.	Num.	Root.
0	0.0000	50	3.6840
1	1.0000	51	7084
2	2509	52	7325
3	4422	53	7563
4	5874	54	7798
5	7100	55	8030
6	8171	56	8259
7	9129	57	8485
8	2.0000	58	8709
9	0801	59	8930
10	2.1544	60	3.9149
11	2240	61	9365
12	2894	62	9579
13	3513	63	9791
14	4101	64	4.0000
15	4602	65	0207
16	5198	66	0412
17	5713	67	0615
18	6207	68	0817
19	6684	69	1016
20	2.7144	70	4.1213
21	7589	71	1408
22	8020	72	1602
23	6439	73	1793
24	8845	74	1983
25	9240	75	2172
26	9625	76	2358
27	3.0000	77	2543
28	0366	78	2727
29	0723	79	2908
30	3.1072	80	4.3089
31	1414	81	3267
32	1748	82	3445
33	2075	83	3621
34	2396	84	3795
35	2711	85	3968
36	3019	86	4140
37	3322	87	4310
38	3620	88	4480
39	3912	89	4647
40	3.4200	90	4.4814
41	4482	91	4979
42	4760	92	5144
43	5084	93	5307
44	5306	94	5468
45	5569	95	5629
46	5830	96	5789
47	6088	97	5947
48	6342	98	6104
49	6596	99	6261
50	6840	100	6416

CUBE ROOTS.

800	0	1	2	3	4	5	6	7	8	9	Differences.
800	90 309	314	320	325	331	336	342	347	352	358	
01	363	369	374	380	385	390	396	401	407	412	
02	417	423	428	434	439	445	450	455	461	466	
03	472	477	482	·488	493	499	504	509	515	520	
04	526	531	536	542	547	553	558	563	569	574	
05	580	585	590	596	601	607	612	617	623	628	
06	634	639	644	650	655	660	666	671	677	682	
07	687	693	698	703	709	714	720	725	730	736	6
08	741	747	752	757	763	768	773	779	784	789	1 1 2 1
09	795	800	806	811	816	822	827	832	838	843	3 2 4 2
810	90 849	854	859	865	870	875	881	886	891	897	5 3 6 4
11	902	907	913	918	924	929	934	940	945	950	7 4 8 5
12	956	961	966	972	977	982	988	993	998	*004	9 5
13	91 009	014	020	025	030	036	041	046	052	057	
14	062	068	073	078	084	089	094	100	105	110	
15	116	121	126	132	137	142	148	153	158	164	
16	169	174	180	185	190	196	201	206	212	217	
17	222	228	233	238	243	249	254	259	265	270	
18	275	281	286	291	297	302	307	312	318	323	
19	328	334	339	344	350	355	360	365	371	376	
820	91 381	387	392	397	403	408	413	418	424	429	
21	434	440	445	450	455	461	466	471	477	482	5
22	487	492	498	503	508	514	519	524	529	535	1 1 2 1
23	540	545	551	556	561	566	572	577	582	587	3 2 4 2
24	593	598	603	609	614	619	624	630	635	640	5 3 6 3
25	645	651	656	661	666	672	677	682	687	693	7 4 8 4
26	698	703	709	714	719	724	730	735	740	745	9 5
27	751	756	761	766	772	777	782	787	793	798	
28	803	808	814	819	824	829	834	840	845	850	
29	855	861	866	871	876	882	887	892	897	903	
830	01 908	913	918	924	929	934	939	944	950	955	
31	960	965	971	976	981	986	991	997	*002	*007	
32	92 012	018	023	028	033	038	044	049	054	059	
33	065	070	075	080	085	091	096	101	106	111	
34	117	122	127	132	137	143	148	153	158	163	
35	169	174	179	184	189	105	200	205	210	215	
36	221	226	231	236	241	247	252	257	262	267	4
37	273	278	283	288	293	298	304	309	314	319	1 0 2 1
38	324	330	335	340	345	350	355	361	366	371	3 1 4 2
39	376	381	387	392	397	402	407	412	418	423	5 2 6 2
840	92 428	433	438	443	449	454	459	464	469	474	7 3 8 3
41	480	485	490	495	500	505	511	516	521	526	9 4
42	531	536	542	547	552	557	562	567	572	578	
43	583	588	593	598	603	609	614	619	624	629	
44	634	639	645	650	655	660	665	670	675	681	
45	686	691	696	701	706	711	716	722	727	732	
46	737	742	747	752	758	763	768	773	778	783	
47	788	793	799	804	809	814	819	824	829	834	
48	840	845	850	855	860	865	870	875	881	886	
49	891	896	901	906	911	916	921	927	932	937	
850	0	1	2	3	4	5	6	7	8	9	Differences.

850	0	1	2	3	4	5	6	7	8	9
850	92 942	947	952	957	962	967	973	978	983	988
51	993	998	*003	*008	*013	*018	*024	*029	*034	*039
52	93 044	049	054	059	064	069	075	080	085	090
53	095	100	105	110	115	120	125	131	136	141
54	146	151	156	161	166	171	176	181	186	192
55	197	202	207	212	217	222	227	232	237	242
56	247	252	258	263	268	273	278	283	288	293
57	298	303	308	313	318	323	328	334	339	344
58	349	354	359	364	369	374	379	384	389	394
59	399	404	409	414	420	425	430	435	440	445
860	93 450	455	460	465	470	475	480	485	490	495
61	500	505*	510	515	520	526	531	536	541	546
62	551	556	561	566	571	576	581	586	591	596
63	601	606	611	616	621	626	631	636	641	646
64	651	656	661	666	671	676	682	687	692	697
65	702	707	712	717	722	727	732	737	742	747
66	752	757	762	767	772	777	782	787	792	797
67	802	807	812	817	822	827	832	837	842	847
68	852	857	862	867	872	877	882	887	892	897
69	902	907	912	917	922	927	932	937	942	947
870	93 952	957	962	967	972	977	982	987	992	997
71	94 002	007	012	017	022	027	032	037	042	047
72	052	057	062	067	072	077	082	086	091	096
73	101	106	111	116	121	126	131	136	141	146
74	151	156	161	166	171	176	181	186	191	196
75	201	206	211	216	221	226	231	236	240	245
76	250	255	260	265	270	275	280	285	290	295
77	300	305	310	315	320	325	330	335	340	345
78	349	354	359	364	369	374	379	384	389	394
79	399	404	409	414	419	424	429	433	438	443
880	94 448	453	458	463	468	473	478	483	488	493
81	498	503	507	512	517	522	527	532	537	542
82	547	552	557	562	567	571	576	581	586	591
83	596	601	606	611	616	621	626	630	635	640
84	645	650	655	660	665	670	675	680	685	689
85	694	699	704	709	714	719	724	729	734	738
86	743	748	753	758	763	768	773	778	783	787
87	792	797	802	807	812	817	822	827	832	836
88	841	846	851	856	861	866	871	876	880	885
89	890	895	900	905	910	915	919	924	929	934
890	94 939	944	949	954	959	963	968	973	978	983
91	988	993	998	*002	*007	*012	*017	*022	*027	*032
92	95 036	041	046	051	056	061	066	071	075	080
93	085	090	095	100	105	109	114	119	124	129
94	134	139	143	148	153	158	163	168	173	177
95	182	187	192	197	202	207	211	216	221	226
96	231	236	240	245	250	255	260	265	270	274
97	279	284	289	294	299	303	308	313	318	323
98	328	332	337	342	347	352	357	361	366	371
99	376	381	386	390	395	400	405	410	415	419
900	0	1	2	3	4	5	6	7	8	9

RECIPROCALS.

Num.	Recip.	Num.	Recip.
.0	∞	5.0	0.2000
.1	10.0000	5.1	1961
.2	5.0000	5.2	1928
.3	3.3333	5.3	1887
.4	2.5000	5.4	1852
.5	2.0000	5.5	1818
.6	1.6667	5.6	1786
.7	4286	5.7	1754
.8	2500	5.8	1724
.9	1111	5.9	1695
1.0	1.0000	6.0	0.1667
1.1	0.9091	6.1	1639
1.2	8333	6.2	1613
1.3	7692	6.3	1587
1.4	7143	6.4	1563
1.5	6667	6.5	1538
1.6	6250	6.6	1515
1.7	5882	6.7	1493
1.8	5556	6.8	1471
1.9	5263	6.9	1449
2.0	0.5000	7.0	0.1429
2.1	4762	7.1	1408
2.2	4545	7.2	1389
2.3	4348	7.3	1370
2.4	4167	7.4	1351
2.5	4000	7.5	1333
2.6	3846	7.6	1316
2.7	3704	7.7	1299
2.8	3571	7.8	1282
2.9	3448	7.9	1266
3.0	0.3333	8.0	0.1250
3.1	3226	8.1	1235
3.2	3125	8.2	1220
3.3	3030	8.3	1205
3.4	2941	8.4	1190
3.5	2857	8.5	1176
3.6	2778	8.6	1163
3.7	2703	8.7	1149
3.8	2632	8.8	1136
3.9	2564	8.9	1124
4.0	0.2500	9.0	0.1111
4.1	2439	9.1	1099
4.2	2381	9.2	1087
4.3	2326	9.3	1075
4.4	2273	9.4	1064
4.5	2222	9.5	1053
4.6	2174	9.6	1042
4.7	2128	9.7	1031
4.8	2083	9.8	1020
4.9	2041	9.9	1010
5.0	0.2000	10.0	0.1000

900	0	1	2	3	4	5	6	7	8	9	Differences.
900	95 424	429	434	439 .	444	448	453	458	463	468	
01	472	477	482	487	492	497	501	506	511	516	
02	521	525	530	535	540	545	550	554	559	564	
03	569	574	578	583	588	593	598	602	607	612	
04	617	622	626	631	636	641	646	650	655	660	
05	665	670	674	679	684	689	694	698	703	708	
06	713	718	722	727	732	737	742	746	751	756	
07	761	766	770	775	780	785	789	794	799	804	
. 08	809	813	818	823 ·	828	832	837	842	847	852	
09	856	861	866	871	875	880	885	890	895	899	
910	95 904	909	914	918	923	928	933	938	942	947	
11	952	957	961	966	971	976	980	985	990	995	5
12	999	*004	*009	*014	*019	*023	*028	*033	*038	*042	1 1
13	96 047	052	057	061	066	071	076	080	085	090	2 1
14	095	099	104	109	114	118	123	128	133	137	3 2 4 2
15	142	147	152	156	161	166	171	175	180	185	5 3
16	190	194	199	204	209	213	218	223	227	232	6 3
17	237	242	246	251	256	261	265	270	275	280	7 4
18	284	289	294	298	303	308	313	317	322	327	8 4 9 5
19	332	336	341	346	350	355	360	365	369	374	
920	96 379	384 .	388	393	398	402	407	412	417	421	
21	426	431	435	440	445	450	454	459	464	468	
22	473	478	483	487	492	497	501	506	511	515	
23	520	525	530	534	539	544	548	553	558	562	
24	567	572	577	581	586	591	595	600	605	609	
25	614	619	624	628	633	638	642	647	652	656	
26	661	666	670	675	680	685	689	694	699	703	
27	708	713	717	722	727	731	736	741	745	750	
28	755	759	764	769	774	778	783	788	792	797	
29	802	806	811	816	820	825	830	834	839	844	
930	96 848	853	858	862	867	872	876	881	886	890	4
31	895	900	904	909	914	918	923	928	932	937	1 0
32	942	946	951	956	960	965	970	974	979	984	2 1
33	988	993	997	*002	*007	*011	*016	*021	*025	*030	3 1
34	97 035	039	044	049	053	058	063	067	072	077	4 2 5 2
35	081	086	090	095	100	104	109	114	118	123	6 2
36	128	132	137	142	146	151	155	160	165	169	7 3
37	174	179	183	188	192	197	202	206	211	216	8 3 9 4
38	220	225	230	234	239	243	248	253	257	262	
39	267	271	276	280	285	290	294	299	304	308	
940	97 313	317	322	327	331	336	340	345	350	354	
41	359	364	368	373	377	382	387	391	396	400	
42	405	410	414	419	424	428	433	437	442	447	
43	451	456	460	465	470	474	479	483	488	493	
44	497	502	506	511	516	520	525	529	534	539	
45	543	548	552	557	562	566	571	575	580	585	
46	589	594	598	603	607	612	617	621	626	630	
47	635	640	644	649	653	658	663	667	672	676	
48	681	685	690	695	699	704	708	713	717	722	
49	727	731	736	740	745	749	754	759	763	768	
950	0	1	2	3	4	5	6	7	8	9	Differences.

950	0	1	2	3	4	5	6	7	8	9
950	97 772	777	782	786	791	795	800	804	809	813
51	818	823	827	832	836	841	845	850	855	859
52	864	868	873	877	882	886	891	896	900	905
53	909	914	918	923	928	932	937	941	946	950
54	955	959	964	968	973	978	982	987	991	996
55	98 000	005	009	014	019	023	028	032	037	041
56	046	050	055	059	064	068	073	078	082	087
57	091	096	100	105	109	114	118	123	127	132
58	137	141	146	150	155	159	164	168	173	177
59	182	186	191	195	200	204	209	214	218	223
960	98 227	232	236	241	245	250	254	259	263	268
61	272	277	281	286	290	295	299	304	308	313
62	318	322	327	331	336	340	345	349	354	358
63	363	367	372	376	381	385	390	394	399	403
64	408	412	417	421	426	430	435	439	444	448
65	453	457	462	466	471	476	480	484	489	493
66	498	502	507	511	516	520	525	529	534	538
67	543	547	552	556	561	565	570	574	579	583
68	588	592	597	601	605	610	614	619	623	628
69	632	637	641	646	650	655	659	664	668	673
970	98 677	682	686	691	695	700	704	709	713	717
71	722	726	731	735	740	744	749	753	758	762
72	767	771	776	780	784	789	793	798	802	807
73	811	816	820	825	829	834	838	843	847	851
74	856	860	865	869	874	878	883	887	892	896
75	900	905	909	914	918	923	927	932	936	941
76	945	949	954	958	963	967	972	976	981	985
77	989	994	998	*003	*007	*012	*016	*021	*025	*029
78	99 034	038	043	047	052	056	061	065	069	074
79	078	083	087	092	096	100	105	109	114	118
980	99 123	127	131	136	140	145	149	154	158	162
81	167	171	176	180	185	189	193	198	202	207
82	211	216	220	224	229	233	238	242	247	251
83	255	260	264	269	273	277	282	286	291	295
84	300	304	308	313	317	322	326	330	335	339
85	344	348	352	357	361	366	370	374	379	383
86	388	392	396	401	405	410	414	419	423	427
87	432	436	441	445	449	454	458	463	467	471
88	476	480	484	489	493	498	502	506	511	515
89	520	524	528	533	537	542	546	550	555	559
990	99 564	568	572	577	581	585	590	594	599	603
91	607	612	616	621	625	629	634	638	642	647
92	651	656	660	664	669	673	677	682	686	691
93	695	699	704	708	712	717	721	726	730	734
94	739	743	747	752	756	760	765	769	774	778
95	782	787	791	795	800	804	808	813	817	822
96	826	830	835	839	843	848	852	856	861	865
97	870	874	878	883	887	891	896	900	904	909
98	913	917	922	926	930	935	939	944	948	952
99	957	961	965	970	974	978	983	987	991	996
1000	0	1	2	3	4	5	6	7	8	9

SINES AND TANGENTS OF SMALL ANGLES.

DEG.	SINE.	TAN.
0°00'	4.68 557	4.68 557
05	557	558
10	557	558
15	557	558
20	557	558
25	557	558
30	557	559
35	557	559
40	557	559
45	556	560
50	556	561
55	556	561
1°00'	4.68 555	4.68 562
05	555	563
10	554	563
15	554	564
20	554	565
25	553	566
30	553	567
35 ·	552	569
40	551	570
45	551	571
50	550	572
55	549	574
2°00'	4.68 549	4.68 575
05	548	577
10	547	578
15	546	580
20	545	582
25	545	583
30	544	585
35	543	587
40	542	589
45	541	591
50	540	593
55	539	595
3°00'	4.68 538	4.68 597
05	537	599
10	535	602
15	534	604
20	533	607
25	532	609
30	530	612
35	529	614
40	528	617
45	526	620
50	525	622
55	524	625
4°00'	522	628

SINES AND TANGENTS OF SMALL ANGLES.

0°	Sines Nat.	Sines Log.	Sines Dif.	Cosines Nat.	Cosines Log.	Cosines Dif.	Tangents Nat.	Tangents Log.	Tangents Dif.	Cotangents Log.	Cotangents Nat.	179°
0'	.00 000	∞	∞	1.0000	0.00 000	.00	.00 000	∞	∞	∞	∞	60'
1	029	6.46 373	501.7	000	000		029	6.46 373	501.7	3.53 627	3437.7	59
2	058	6.76 476	298.5	000	000		058	6.76 476	298.5	3.23 524	1718.9	58
3	087	6.94 085	208.2	000	000		087	6.94 085	208.2	3.05 915	1145.9	57
4	116	7.06 579	161.5	000	000		116	7.06 579	161.5	2.93 421	859.44	56
5	145	7.16 270	132.0	000	000		145	7.16 270	132.0	2.83 730	687.55	55
6	175	7.24 188	111.6	000	000		175	7.24 188	111.6	2.75 812	572.96	54
7	204	7.30 882	96.7	000	000		204	7.30 882	96.7	2.69 118	491.11	53
8	233	7.36 682	85.8	000	000		238	7.36 682	85.3	2.63 318	429.72	52
9	262	7.41 797	76.8	000	000		262	7.41 797	76.8	2.58 203	381.97	51
10	.00 291	7.46 373	69.0	1.0000	0.00 000	.00	.00 291	7.46 373	69.0	2.53 627	343.77	50
11	320	7.50 512	63.0	.99 999	000		320	7.50 512	63.0	2.49 488	312.52	49
12	349	7.54 291	57.9	999	000		849	7.54 291	57.9	2.45 709	286.48	48
13	378	7.57 767	53.6	999	000		378	7.57 767	53.7	2.42 233	264.44	47
14	407	7.60 985	50.0	999	000		407	7.60 986	49.9	2.39 014	245.55	46
15	436	7.63 982	46.7	999	000		436	7.63 982	46.7	2.36 018	229.18	45
16	465	7.66 784	43.9	999	000	.02	465	7.66 785	43.9	2.33 215	214.86	44
17	495	7.69 417	41.4	999	9.99 999	.00	495	7.69 418	41.4	2.30 582	202.22	43
18	524	7.71 900	39.1	999	999		524	7.71 900	80.1	2.28 100	190.98	42
19	553	7.74 248	37.1	998	999		553	7.74 248	37.1	2.25 752	180.93	41
20	.00 582	7.76 475	35.8	.99 998	9.99 999	.00	.00 582	7.76 476	35.8	2.23 524	171.89	40
21	611	7.78 594	33.7	998	999		611	7.78 595	33.7	2.21 405	168.70	39
22	640	7.80 615	32.2	998	999		640	7.80 615	32.2	2.19 385	156.26	38
23	669	7.82 545	30.6	998	999		669	7.82 546	30.6	2.17 454	149.47	37
24	698	7.84 393	29.6	998	999		698	7.84 394	29.6	2.15 606	143.24	36
25	727	7.86 166	28.4	997	999		727	7.86 167	28.4	2.13 833	137.51	35
26	756	7.87 870	27.8	997	999		756	7.87 871	27.8	2.12 129	182.92	·34
27	785	7.89 509	26.8	997	999		785	7.89 510	26.8	2.10 490	127.82	33
28	814	7.91 088	25.4	997	999	.02	0815	7.91 089	25.4	2.08 911	122.77	32
29	644	7.92 612	24.5	996	998	.00	844	7.92 613	24.6	2.07 387	118.54	31
30	.00 673	7.94 084	23.7	.99 996	9.99 998	.00	.00 873	7.94 086	23.7	2.05 914	114.59	30
31	902	7.95 508	28.0	996	998		902	7.95 510	23.0	2.04 490	110.89	29
32	981	7.96 887	22.8	996	998		981	7.96 889	22.3	2.03 111	107.48	28
33	960	7.98 223	21.6	995	998		960	7.98 225	21.6	2.01 775	104.17	27
34	989	7.99 520	21.0	995	998		989	7.99 522	21.0	2.00 478	101.11	26
35	.01 018	8.00 779	20.4	995	998		.01 018	8.00 781	20.4	1.99 219	98.218	25
36	047	8.02 002	19.8	995	998	.02	047	8.02 004	19.8	1.97 996	95.489	24
37	076	8.03 192	19.8	994	997	.00	076	8.03 194	19.8	1.96 806	92.908	23
38	105	8.04 350	18.8	994	997		105	8.04 353	18.8	1.95 647	90.463	22
39	134	8.05 478	18.3	994	997		134	8.05 481	18.8	1.94 519	88.144	21
40	.01 164	8.06 578	17.9	.99 993	9.99 997	.00	.01 164	8.06 581	17.9	1.93 419	85.940	20
41	193	8.07 650	17.4	993	997		193	8.07 653	17.5	1.92 347	83.844	19
42	222	8.08 696	17.0	993	997		222	8.08 700	17.0	1.91 300	81.847	18
43	251	8.09 718	16.7	992	997	.02	251	8.09 722	16.6	1.90 278	79.943	17
44	280	8.10 717	16.3	992	996	.00	280	8.10 720	16.3	1.89 280	78.126	16
45	309	8.11 693	15.9	991	996		309	8.11 696	15.9	1.88 304	76.890	15
46	338	8.12 647	15.6	991	996		338	8.12 651	15.6	1.87 349	74.729	14
47	367	8.13 581	15.2	991	996		367	8.13 585	15.3	1.86 415	73.139	13
48	396	8.14 495	14.9	990	996		396	8.14 500	14.9	1.85 500	71.615	12
49	425	8.15 391	14.6	990	996	.02	425	8.15 395	14.6	1.84 605	70.153	11
50	.01 454	8.16 268	14.3	.99 989	9.99 995	.00	.01 455	8.16 273	14.3	1.83 727	68.750	10
51	483	8.17 128	14.1	989	995		484	8.17 133	14.1	1.82 867	67.402	9
52	513	971	13.8	989	995		513	976	13.8	024	66.105	8
53	542	8.18 798	13.5	988	995		542	8.18 804	13.5	1.81 196	64.858	7
54	571	8.19 610	13.3	988	995	.02	571	8.19 616	13.3	1.80 384	63.657	6
55	600	8.20 407	13.0	987	994	.00	600	8.20 413	13.0	1.79 587	62.499	5
56	629	8.21 189	12.5	987	994		629	8.21 195	12.6	1.78 805	61.383	4
57	658	958	12.6	986	·994		658	964	12.6	036	60.806	3
58	687	8.22 713	12.4	986	994		687	8.22 720	12.4	1.77 280	59.266	2
59	716	8.23 456	12.2	985	994	.02	716	8.23 462	12.2	1.76 538	58.261	1
60	.01 745	8.24 186		.99 985	9.99 993		.01 746	8.24 192		1.75 808	57.290	0

90°	Nat.	Log.	Dif.	Nat.	Log.	Dif.	Nat.	Log.	Dif.	Log.	Nat.	89°
	Cosines.			Sines.			Cotangents.			Tangents.		

1°	Sines. Nat.	Log.	Dif.	Cosines. Nat.	Log.	Dif.	Tangents. Nat.	Log.	Dif.	Cotangents. Log.	Nat.	178°
0'	.01 745	8.24 186	12.0	.99 985	9.99 993	.00	.01 746	8.24 192	12.0	1.75 808	57.290	60'
1	774	903	11.9	984	993		775	910	11.8	090	56.351	59
2	803	8.25 609	11.6	984	993		804	8.25 616	11.6	1.74 384	55.442	58
3	832	8.26 304	11.4	983	993	.02	833	8.26 312	11.4	1.73 688	54.561	57
4	862	988	11.2	983	992	.00	862	996	11.2	004	53.709	56
5	891	8.27 661	11.1	982	992		891	8.27 669	11.1	1.72 331	52.882	55
6	920	8.28 324	10.0	982	992		920	8.28 332	10.9	1.71 668	.081	54
7	949	977	10.7	981	992		949	986	10.7	014	51.303	53
8	978	8.29 621	10.6	980	992	.02	978	8.29 629	10.6	1.70 371	50.549	52
9	.02 007	8.30 255	10.4	980	991	.00	.02 007	8.30 263	10.4	1.69 737	49.816	51
10	.02 036	8.30 879	10.3	.99 979	9.99 991	.00	.02 036	8.30 888	10.3	1.69 112	49.104	50
11	065	8.31 495	10.1	979	991	.02	066	8.31 505	10.1	1.68 495	48.412	49
12	094	8.32 103	10.0	978	990	.00	095	8.32 112	10.0	1.67 888	47.740	48
13	123	702	9.8	977	990		124	711	9.9	289	.085	47
14	152	8.33 292	9.7	977	990		153	8.33 302	9.7	1.66 698	46.449	46
15	181	875	9.6	976	990	.02	182	886	9.6	114	45.829	45
16	211	8.34 450	9.5	976	989	.00	211	8.34 461	9.5	1.65 539	.226	44
17	240	8.35 018	9.3	975	989		240	8.35 029	9.4	1.64 971	44.699	43
18	269	578	9.2	974	989		269	590	9.2	410	.066	42
19	298	8.36 131	9.1	974	989	.02	298	8.36 143	9.1	1.63 857	43.508	41
20	.02 327	8.36 678	9.0	.99 973	9.99 988	.00	.02 328	8.36 689	9.0	1.63 311	42.964	40
21	356	8.37 217	8.9	972	988		357	8.37 229	8.9	1.62 771	.438	39
22	385	750	8.8	972	988	.02	386	762	8.8	238	41.916	38
23	414	8.38 276	8.7	971	987	.00	415	8.38 289	8.7	1.61 711	.411	37
24	443	796	8.6	970	987		444	809	8.6	191	40.917	36
25	472	8.39 310	8.5	969	987	.02	473	8.39 323	8.5	1.60 677	.436	35
26	501	818	8.4	969	986	.00	502	832	8.4	168	39.965	34
27	530	8.40 320	8.8	968	986		531	8.40 334	8.3	1.59 666	.506	33
28	560	816	8.2	967	986	.02	560	830	8.2	170	.057	32
29	589	8.41 307	8.1	966	985	.00	589	8.41 321	8.1	1.58 679	38.618	31
30	.02 618	8.41 792	8.0	.99 966	9.99 985	.00	.02 619	8.41 807	8.0	1.58 193	88.168	30
31	647	8.42 272	7.9	965	985	.02	648	8.42 287	7.9	1.57 713	87.769	29
32	676	746	7.8	964	984	.00	677	762	7.8	238	.386	28
33	705	8.43 216	7.7	963	984		706	8.43 232	7.7	1.56 768	86.956	27
34	734	680		963	984	.02	735	696		304	.563	26
35	763	8.44 139	7.6	962	983	.00	764	8.44 156	7.6	1.55 844	.178	25
36	792	594	7.5	961	983		793	611	7.5	389	85.801	24
37	821	8.45 044	7.4	960	983	.02	822	8.45 061	7.4	1.54 939	.481	23
38	850	489		959	982	.00	851	507		493	.070	22
39	879	930	7.3	959	982		881	948	7.3	052	84.715	21
40	.02 908	8.46 366	7.2	.99 958	9.99 982	.02	.02 910	8.46 385	7.2	1.53 615	84.868	20
41	938	799	7.1	957	981	.00	939	817	7.1	183	.027	19
42	967	8.47 226		956	981		968	8.47 245		1.52 755	83.694	18
43	996	650	7.0	955	981	.02	997	669	7.0	331	.366	17
44	.03 025	8.48 069	6.9	954	980	.00	.03 026	8.48 089	6.9	1.51 911	.045	16
45	.03 054	485		953	980	.02	055	505		495	82.780	15
46	083	896	6.8	952	979	.00	084	917	6.8	083	.421	14
47	112	8.49 304	6.7	952	979		114	8.49 325	6.7	1.50 675	.118	13
48	141	708		951	979	.02	143	729		271	81.821	12
49	170	8.50 108	6.6	950	978		172	8.50 130	6.6	1.49 870	.528	11
50	.03 199	8.50 504	6.6	.99 949	9.99 978	.02	.03 201	8.50 527	6.6	1.49 473	81.242	10
51	228	897	6.5	949	977	.00	230	920	6.5	080	30.960	9
52	257	8.51 287	6.4	947	977		259	8.51 310	6.4	1.48 690	.663	8
53	286	673		946	977	.02	288	696		304	.412	7
54	316	8.52 055	6.3	945	976	.00	317	8.52 079	6.3	1.47 921	.145	6
55	345	434		944	976	.02	346	459		541	29.882	5
56	374	810	6.2	943	975	.00	376	835	6.2	165	.624	4
57	403	8.53 183		942	975	.02	405	8.53 208		1.46 792	.371	3
58	432	552	6.1	941	974	.00	434	578	6.1	422	.122	2
59	461	919		940	974		463	945		055	28.977	1
60	.03 490	8.54 282		.99 939	9.99 974		.03 492	8.54 308		1.45 692	28.636	0

91°	Nat.	Log.	Dif.	Nat.	Log.	Dif.	Nat.	Log.	Dif.	Log.	Nat.	88°
		Cosines.			Sines.			Cotangents.			Tangents.	

2°	SINES			COSINES			TANGENTS			COTANGENTS		177°
'	Nat.	Log.	Dif.	Nat.	Log.	Dif.	Nat.	Log.	Dif.	Log.	Nat.	
0'	.03 490	8.54 282	6.0	.99 939	9.99 974	.02	.03 492	8.54 308	6.0	1.45 692	28.636	60'
1	519	642		938	973	.00	521	669		331	.809	59
2	548	999	5.9	937	973	.02	550	8.55 027	5.9	1.44 973	.166	58
3	577	8.55 354		936	972	.00	579	382		618	27.987	57
4	606	705	5.8	935	972	.02	609	734	5.8	266	.712	56
5	635	8.56 054		934	971	.00	638	8.56 083		1.43 917	.490	55
6	664	400	5.7	933	971	.02	667	429	5.7	571	.271	54
7	693	743		932	970	.00	696	773		227	.057	53
8	723	8.57 084	5.6	931	970	.02	725	8.57 114	5.6	1.42 886	26.845	52
9	752	421		930	969	.00	754	452		548	.637	51
10	.03 781	8.57 757	5.5	.99 929	9.99 969	.02	.03 783	8.57 788	5.6	1.42 212	26.432	50
11	810	8.58 089		927	968	.00	812	8.58 121	5.5	1.41 879	.230	49
12	839	419		926	968	.02	842	451		549	.031	48
13	868	747	5.4	925	967	.00	871	779	5.4	221	25.835	47
14	897	8.59 072		924	967		900	8.59 105		1.40 895	.642	46
15	926	395	5.3	923	967	.02	929	428		572	.452	45
16	955	715		922	966	.00	958	749	5.3	251	.264	44
17	984	8.60 033		921	966	.02	987	8.60 068		1.39 932	.080	43
18	.04 013	349	5.2	919	965		.04 016	384	5.2	616	24.898	42
19	042	662		918	964	.00	046	698		302	.719	41
20	.04 071	8.60 973	5.2	.99 917	9.99 964	.02	.04 075	8.61 009	5.2	1.38 991	24.542	40
21	100	8.61 282	5.1	916	963	.00	104	319	5.1	681	.368	39
22	129	589		915	963	.02	133	626		374	.196	38
23	159	894	5.0	913	962	.02	162	931		069	.026	37
24	188	8.62 196		912	962	.02	191	8.62 234	5.0	1.37 766	23.859	36
25	217	497		911	961	.00	220	535		465	.695	35
26	246	795	4.9	910	961	.02	250	834		166	.592	34
27	275	8.63 091		909	960	.00	279	8.63 131	4.9	1.36 869	.372	33
28	304	385		907	960	.02	308	426		574	.214	32
29	333	678	4.8	906	959	.00	337	718		282	.058	31
30	.04 362	8.63 968	4.8	.99 905	9.99 959	.02	.04 366	8.64 009	4.8	1.35 991	22.904	30
31	391	8.64 256		904	958	.00	395	298		702	.752	29
32	420	543	4.7	902	958	.02	424	585		415	.602	28
33	449	827		901	957		454	870	4.7	130	.454	27
34	478	8.65 110		900	956	.00	483	8.65 154		1.34 846	.308	26
35	507	391		898	956	.02	512	435		565	.164	25
36	536	670	4.6	897	955	.00	541	715	4.6	285	.022	24
37	565	947		896	955	.02	570	993		007	21.881	23
38	594	8.66 223		894	954	.00	599	8.66 269		1.33 731	.748	22
39	623	497	4.5	893	954	.02	628	543		457	.606	21
40	.04 653	8.66 769	4.5	.99 892	9.99 953	.02	.04 658	8.66 816	4.5	1.33 184	21.470	20
41	682	8.67 039		890	952	.00	687	8.67 087		1.32 913	.337	19
42	711	308		889	952	.02	716	356		644	.205	18
43	740	575	4.4	888	951	.00	745	624	4.4	376	.075	17
44	769	841		886	951	.02	774	890		110	20.946	16
45	798	8.68 104		885	950		803	8.68 154		1.31 846	.819	15
46	827	367	4.3	883	949	.00	832	417		583	.698	14
47	856	627		882	949	.02	862	678	4.3	322	.569	13
48	885	886		881	948	.00	891	938		062	.446	12
49	914	8.69 144		879	948	.02	920	8.69 196		1.30 804	.325	11
50	.04 943	8.69 400	4.2	.99 878	9.99 947	.02	.04 949	8.69 453	4.3	1.30 547	20.206	10
51	972	654		876	946	.00	978	708	4.2	292	.087	9
52	.05 001	907		875	946	.02	.05 007	962		038	19.970	8
53	030	8.70 159		873	945		037	8.70 214		1.29 786	.855	7
54	059	409		872	944	.00	066	465		535	.740	6
55	088	658	4.1	870	944	.02	095	714	4.1	286	.027	5
56	117	905		869	943		124	962		038	.516	4
57	146	8.71 151		867	942	.00	153	8.71 208		1.28 792	.405	3
58	175	395		866	942	ʼ.02	182	453		547	.296	2
59	205	638	4.0	864	941		212	697		303	.188	1
60	.05 234	8.71 880		.99 863	9.99 940		.05 241	8.71 940		1.28 060	19.081	0

92°	Nat.	Log.	Dif.	Nat.	Log.	Dif.	Nat.	Log.	Dif.	Log.	Nat.	87°
		COSINES.			SINES.			COTANGENTS.		TANGENTS.		

3°	SINES.			COSINES.			TANGENTS.			COTANGENTS		176°
	Nat.	Log.	Dif.	Nat.	Log.	Dif.	Nat.	Log.	Dif.	Log.	Nat.	
0'	.05 284	8.71 880	4.0	.99 863	9.99 940	.00	.05 241	8.71 940	4.0	1.28 060	19.081	60'
1	263	8.72 120		861	940	.02	270	8.72 181		1.27 819	18.976	59
2	292	359		860	939		299	420		580	.871	58
3	321	597		858	938	.00	329	659		341	.768	57
4	350	834	3.9	857	933	.02	357	896	3.9	104	.666	56
5	379	8.73 069		855	937		387	8.73 132		1.26 868	.564	55
6	408	303		854	936	.00	416	366		634	.464	54
7	437	535		852	936	.02	445	600		400	.366	53
8	466	767	3.8	851	935		474	832		168	.268	52
9	495	997		849	934	.00	503	8.74 063	3.8	1.25 937	.171	51
10	.05 524	8.74 226	3.8	.99 847	9.99 934	.02	.05 583	8.74 292	3.8	1.25 708	18.075	50
11	553	454		846	933		562	521		479	17.980	49
12	582	680		844	932	.00	591	748		252	.886	48
13	611	906	3.7	842	932	.02	620	974		026	.798	47
14	640	8.75 130		841	931		649	8.75 199	3.7	1.24 801	.702	46
15	669	353		839	930		678	423		577	.611	45
16	698	575		838	929	.00	708	645		355	.521	44
17	727	795		836	929	.02	737	867		133	.431	43
18	756	8.76 015		834	928		766	8.76 087		1.23 913	.343	42
19	785	234	3.6	833	927		795	306		694	.256	41
20	.05 814	8.76 451	3.6	.99 831	9.99 926	.00	.05 824	8.76 525	3.6	1.23 475	17.169	40
21	844	667		829	926	.02	854	742		258	.084	39
22	873	883		827	925		883	958		042	16.999	38
23	902	8.77 097		826	924		912	8.77 173		1.22 827	.915	37
24	931	310	3.5	824	923	.00	941	387		613	.832	36
25	960	522		822	923	.02	970	600	3.5	400	.750	35
26	989	733		821	922		.05 999	811		189	.668	34
27	.06 018	943		819	921		.06 029	8.78 022		1.21 978	.587	33
28	047	8.78 152		817	920	.00	058	232		768	.507	32
29	076	360		815	920	.02	087	441		559	.428	31
30	.06 105	8.78 568	3.4	.99 813	9.99 919	.02	.06 116	8.78 649	3.4	1.21 351	16.350	30
31	134	774		812	918		145	855		145	.272	29
32	163	979		810	917	.00	175	8.79 061		1.20 939	.195	28
33	192	8.79 183		808	917	.02	204	266		734	.119	27
34	221	386		806	916		233	470		530	.043	26
35	250	588		804	915		262	673		327	15.969	25
36	279	789		808	914		291	875		125	.895	24
37	308	990	3.3	801	913	.00	321	8.80 076		1.19 924	.821	23
38	337	8.80 189		799	913	.02	350	277	3.3	723	.748	22
39	366	388		797	912		379	476		524	.676	21
40	.06 395	8.80 585	3.3	.99 795	9.99 911	.02	.06 408	8.80 674	3.3	1.19 326	15.605	20
41	424	782		793	910		438	872		128	.534	19
42	453	978		792	909	.00	467	8.81 068		1.18 932	.464	18
43	482	8.81 173	3.2	790	909	.02	496	264		736	.394	17
44	511	367		788	908		525	459	3.2	541	.325	16
45	540	560		786	907		554	653		347	.257	15
46	569	752		784	906		584	846		154	.189	14
47	598	944		782	905		613	8.82 038		1.17 962	.122	13
48	627	8.82 134		780	904	.00	642	230		770	.056	12
49	656	324		778	904	.02	671	420		580	14.990	11
50	.06 685	8.82 513	3.1	.99 776	9.99 903	.02	.06 700	8.82 610	3.2	1.17 390	14.924	10
51	714	701		774	902		730	799	3.1	201	.860	9
52	743	888		772	901		759	987		013	.795	8
53	772	8.83 075		770	900		788	8.83 175		1.16 825	.732	7
54	802	261		768	899		817	361		639	.669	6
55	831	446		766	898	.00	847	547		453	.606	5
56	860	630		764	898	.02	876	732		268	.544	4
57	889	813		762	897		905	916		084	.482	3
58	918	996	3.0	760	896		934	8.84 100	3.0	1.15 900	.421	2
59	947	8.84 177		758	895		963	282		718	.361	1
60	.06 976	8.84 358		.99 756	9.99 894		.06 993	8.84 464		1.15 536	14.301	0
93°	Nat.	Log.	Dif.	Nat.	Log.	Dif.	Nat.	Log.	Dif.	Log.	Nat.	86°
		COSINES.			SINES.			COTANGENTS.			TANGENTS.	

4°	SINES			COSINES			TANGENTS			COTANGENTS		175°
	Nat.	Log.	Dif.	Nat.	Log.	Dif.	Nat.	Log.	Dif.	Log.	Nat.	
0′	.06 976	8.84 358	3.0	.99 756	9.99 894	.02	.06 993	8.84 464	3.0	1.15 536	14.301	60′
1	.07 005	539		754	893		.07 022	646		354	.241	59
2	034	718		752	892		051	826		174	.182	58
3	063	897		750	891	.00	080	8.85 006		1.14 994	.124	57
4	092	8.85 075		748	891	.02	110	185		815	.065	56
5	121	252		746	890		139	363		637	.008	55
6	150	429	2.9	744	889		168	540		460	13.951	54
7	179	605		742	888		197	717	2.9	283	.894	53
8	208	780		740	887		227	893		107	.838	52
9	237	955		738	886		256	8.86 069		1.13 931	.782	51
10	.07 266	8.86 128	2.9	.99 736	9.99 885	.02	.07 285	8.86 243	2.9	1.13 757	13.727	50
11	295	301		734	884		314	417		583	.672	49
12	324	474		731	883		344	591		409	.617	48
13	353	645		729	882		373	763		237	.563	47
14	382	816		727	881		402	935		065	.510	46
15	411	987	2.8	725	880		431	8.87 106		1.12 894	.457	45
16	440	8.87 156		723	879	.00	461	277	2.8	723	.404	44
17	469	325		721	879	.02	490	447		553	.352	43
18	498	494		719	878		519	616		384	.300	42
19	527	661		716	877		548	785		215	.248	41
20	.07 556	8.87 829	2.8	.99 714	9.99 876	.02	.07 578	8.87 953	2.8	1.12 047	13.197	40
21	585	995		712	875		607	8.88 120		1.11 880	.146	39
22	614	8.88 161		710	874		636	287		713	.096	38
23	643	326	2.7	708	873		665	453		547	.046	37
24	672	490		705	872		695	618		382	12.996	36
25	701	654		703	871		724	783		217	.947	35
26	730	817		701	870		753	948	2.7	052	.898	34
27	759	980		699	869		782	8.89 111		1.10 889	.850	33
28	788	8.89 142		696	868		812	274		726	.801	32
29	817	304		694	867		841	437		563	.754	31
30	.07 846	8.89 464	2.7	.99 692	9.99 866	.02	.07 870	8.89 598	2.7	1.10 402	12.706	30
31	875	625		689	865		899	760		240	.659	29
32	904	784		687	864		929	920		080	.612	28
33	933	943		685	863		958	8.90 080		1.09 920	.566	27
34	962	8.90 102	2.6	683	862		987	240		760	.520	26
35	991	260		680	861		.08 017	399	2.6	601	.474	25
36	.08 020	417		678	860		046	557		443	.429	24
37	049	574		676	859		075	715		285	.384	23
38	078	730		673	858		104	872		128	.339	22
39	107	885		671	857		134	8.91 029		1.08 971	.295	21
40	.08 136	8.91 040	2.6	.99 668	9.99 856	.02	.08 163	8.91 185	2.6	1.08 815	12.251	20
41	165	195		666	855		192	340		660	.207	19
42	194	349		664	854		221	495		505	.163	18
43	223	502		661	853		251	650		350	.120	17
44	252	655	2.5	659	852		280	803		197	.077	16
45	281	807		657	851		309	957		043	.035	15
46	310	959		654	850	.03	339	8.92 110		1.07 890	11.992	14
47	339	8.92 110		652	848	.02	368	262	2.5	738	.950	13
48	368	261		649	847		397	414		586	.909	12
49	397	411		647	846		427	565		435	.867	11
50	.08 426	8.92 561	2.5	.99 644	9.99 845	.02	.08 456	8.92 716	2.5	1.07 284	11.826	10
51	455	710		642	844		485	866		134	.785	9
52	484	859		639	843		514	8.93 016		1.06 984	.745	8
53	513	8.93 007		637	842		544	165		835	.705	7
54	542	154		635	841		573	313		687	.664	6
55	571	301		632	840		602	462		538	.625	5
56	600	448	2.4	630	839		632	609		391	.585	4
57	629	594		627	838		661	756		244	.546	3
58	658	740		625	837		690	903	2.4	097	.507	2
59	687	885		622	836	.03	720	8.94 049		1.05 951	.469	1
60	.08 716	8.94 030		.99 619	9.99 834		.08 749	8.94 195		1.05 805	11.430	0
94°	Nat.	Log.	Dif.	Nat.	Log.	Dif.	Nat.	Log.	Dif.	Log.	Nat.	85°
		COSINES.			SINES.			COTANGENTS.			TANGENTS.	

5°	Sines.			Cosines.			Tangents.			Cotangents.		174°
	Nat.	Log.	Dif.	Nat.	Log.	Dif.	Nat.	Log.	Dif.	Log.	Nat.	
0'	.08 716	8.94 030	2.4	.99 619	9.99 834	.02	.08 749	8.04 195	2.4	1.05 805	11.480	60'
1	745	174		617	833		778	340		660	.892	59
2	774	317		614	832		807	485		515	.354	58
3	603	461		612	831		887	630		370	.816	57
4	831	603		609	830		866	773		227	.279	56
5	860	746		607	829		895	917		083	.242	55
6	889	887		604	828		925	8.95 060		1.04 940	.205	54
7	919	8.95 029		602	827	.03	954	202		798	.168	53
8	947	170	2.8	599	825	.02	983	344		656	.132	52
9	976	310		596	824		.09 013	486		514	.095	51
10	.09 005	8.95 450	2.8	.99 594	9.99 823	.02	.09 042	8.95 627	2.3	1.04 373	11.059	50
11	084	589		591	822		071	767	2.4	233	.024	49
12	063	728		588	821		101	908	2.3	092	10.988	48
13	092	867		586	820		130	8.96 047		1.03 953	.938	47
14	121	8.96 005		583	819	.03	159	187		813	.918	46
15	150	143		580	817	.02	189	325		675	.883	45
16	179	280		578	816		218	464		536	.848	44
17	208	417		575	815		247	602		398	.814	43
18	237	553		572	814		277	739		261	.780	42
19	266	689		570	813		306	877		123	.746	41
20	.09 295	8.96 825	2.8	.99 567	9.99 812	.03	.09 335	8.97 013	2.8	1.02 987	10.712	40
21	324	960		564	810	.02	365	150		850	.678	39
22	353	8.97 095	2.2	562	809		394	285		715	.645	38
23	382	229		559	808		423	421		579	.612	37
24	411	363		556	807		453	556		444	.579	36
25	440	496		553	806	03	482	691	2.2	309	.546	35
26	469	629		551	804	.02	511	825		175	.514	34
27	498	762		548	803		541	959		041	.481	33
28	527	894		545	802		570	8.98 092		1.01 908	.449	32
29	556	8.98 026		542	801		600	225		775	.417	31
30	.09 585	8.98 157	2.2	.99 540	9.99 800	.03	.09 629	8.98 358	2.2	1.01 642	10.385	30
31	614	288		537	798	.02	658	490		510	.354	29
32	642	419		534	797		688	622		378	.322	28
33	671	549		531	796		717	753		247	.291	27
34	700	679		528	795	.03	746	884		116	.260	26
35	729	808		526	793	.02	776	8.99 015		1.00 985	.229	25
36	758	937		523	792		805	145		855	.199	24
37	787	8.99 066	2.1	520	791		834	275		725	.168	23
38	816	194		517	790	.03	864	405		595	.138	22
39	845	322		514	788	.02	893	534	2.1	466	.108	21
40	.09 874	8.99 450	2.1	.99 511	9.99 787	.02	.09 923	8.99 662	2.2	1.00 338	10.078	20
41	903	577		508	786		952	791	2.1	209	.048	19
42	932	704		506	785	.03	981	919		081	.019	18
43	961	830		503	783	.02	.10 011	9.00 046		0.99 954	9.9893	17
44	990	956		500	782		040	174		826	801	16
45	.10 019	9.00 082		497	781		069	301		699	810	15
46	048	207		494	780	.03	099	427		573	021	14
47	077	332		491	778	.02	128	553		447	9.8734	13
48	106	456		488	777		158	679		321	448	12
49	135	581		485	776		187	805		195	164	11
50	.10 164	9.00 704	2.1	.99 482	9.99 775	.08	.10 216	9.00 930	2.1	0.99 070	9.7682	10
51	192	828		479	773	.02	246	9.01 055		0.98 945	601	9
52	221	951		476	772		275	179		821	322	8
53	250	9.01 074	2.0	473	771	.08	305	303		697	044	7
54	279	196		470	769	.02	334	427		573	9.6768	6
55	308	318		467	768		363	550		450	409	5
56	337	440		464	767	.03	393	673		327	220	4
57	366	561		461	765	.02	422	796	2.0	204	9.5949	3
58	395	682		459	764		452	918		082	679	2
59	424	803		455	763	.03	481	9.02 040		0.97 960	411	1
60	.10 453	9.01 923		.99 452	9.99 761		.10 510	9.02 162		0.97 838	9.5144	0
95°	Nat.	Log.	Dif.	Nat.	Log.	Dif.	Nat.	Log.	Dif.	Log.	Nat.	84°
		Cosines.			Sines.			Cotangents.		Tangents.		

6°	Sines.			Cosines.			Tangents.			Cotangents.		173°
	Nat.	Log.	Dif.	Nat.	Log.	Dif.	Nat.	Log.	Dif.	Log.	Nat.	
0′	.10 453	9.01 923	2.0	.99 452	9.99 761	.02	.10 510	9.02 162	2.0	0.97 838	9.5144	60′
1	482	9.02 043		449	760		540	283		717	9.4878	59
2	511	163		446	759	.03	569	404		596	614	58
3	540	283		443	757	.03	599	525		475	352	57
4	569	402		440	756		628	645		355	090	56
5	597	520		437	755	.03	657	766		234	9.3831	55
6	626	639		434	753	.02	687	885		115	572	54
7	655	757		431	752		716	9.03 005		0.96 995	315	53
8	684	874		428	751	.03	746	124		876	090	52
9	713	992		424	749	.02	775	242		758	9.2806	51
10	.10 742	9.03 109	2.0	.99 421	9.99 748	.02	.10 805	9.03 361	2.0	0.96 639	9.2253	50
11	771	226	1.9	418	747	.03	834	479		521	302	49
12	800	342		415	745	.02	863	597		403	052	48
13	829	458		412	744	.03	893	714		286	9.1603	47
14	858	574		409	742	.02	922	832	1.9	168	555	46
15	887	690		406	741		952	948	2.0	052	309	45
16	916	805		402	740	.03	981	9.04 065	1.9	0.95 935	065	44
17	945	920		399	738	.02	.11 011	181		819	9.0821	43
18	973	9.04 034		396	737		040	297		703	579	42
19	.11 002	149		393	736	.03	070	413		587	838	41
20	.11 031	9.04 262	1.9	.99 390	9.99 734	.03	.11 099	9.04 528	1.9	0.95 472	9.0098	40
21	060	376		386	733	.03	128	643		357	8.9560	39
22	089	490		383	731	.02	158	758		242	628	38
23	118	603		380	730	.03	187	873		127	387	37
24	147	715		377	728	.02	217	987		013	152	36
25	176	828		374	727		246	9.05 101		0.94 899	8.6919	35
26	205	940		370	726	.03	276	214		786	686	34
27	234	9.05 052		367	724	.02	305	328		672	455	33
28	263	164		364	723	.03	335	441		559	225	32
29	291	275		360	721	.02	364	553		447	8.7996	31
30	.11 320	9.05 386	1.9	.99 357	9.99 720	.03	.11 394	9.05 666	1.9	0.94 334	8.7769	30
31	349	497	1.8	354	718	.02	423	778		222	542	29
32	378	607		351	717		452	890		110	317	28
33	407	717		347	716	.03	482	9.06 002		0.93 998	093	27
34	436	827		344	714	.02	511	113		887	8.6870	26
35	465	937		341	713	.03	541	224		776	648	25
36	494	9.06 046		337	711	.02	570	335	1.8	665	427	24
37	523	155		334	710	.03	600	445	1.9	555	208	23
38	552	264		331	708	.02	629	556	1.8	444	8.5989	22
39	580	372		327	707	.03	659	666		334	772	21
40	.11 609	9.06 481	1.8	.99 324	9.99 705	.02	.11 688	9.06 775	1.8	0.93 225	8.5555	20
41	639	589		320	704	.03	718	885		115	340	19
42	667	696		317	702	.02	747	994		006	126	18
43	696	804		314	701	.03	777	9.07 103		0.92 897	8.4913	17
44	725	911		310	699	.02	806	211		789	701	16
45	754	9.07 018		307	698	.03	836	320		680	490	15
46	783	124		303	696	.02	865	428		572	280	14
47	812	231		300	695	.03	895	536		464	071	13
48	840	337		297	693	.02	924	643		357	8.3863	12
49	869	442		293	692	.03	954	751		249	656	11
50	.11 898	9.07 548	1.8	.99 290	9.99 691	.02	.11 963	9.07 858	1.8	0.92 142	8.8450	10
51	927	653		286	689	.03	.12 013	964		036	245	9
52	956	758		283	687	.02	042	9.08 071		0.91 929	041	8
53	985	863		279	686	.03	072	177		823	8.2838	7
54	.12 014	968	1.7	276	684	.02	101	283		717	636	6
55	043	9.08 072		272	683	.03	131	389		611	434	5
56	071	176		269	681	.02	160	495		505	234	4
57	100	280		265	680	.03	190	600		400	035	3
58	129	383		262	678	.02	219	705		295	8.1837	2
59	158	486		258	677	.03	249	810	1.7	190	640	1
60	.12 187	9.08 589		.99 255	9.99 675		.12 278	9.08 914		0.91 086	8.1443	0

96°	Nat.	Log.	Dif.	Nat.	Log.	Dif.	Nat.	Log.	Dif.	Log.	Nat.	83°
		Cosines.			Sines.			Cotangents.			Tangents.	

7°	Sines Nat.	Sines Log.	Sines Dif.	Cosines Nat.	Cosines Log.	Cosines Dif.	Tangents Nat.	Tangents Log.	Tangents Dif.	Cotangents Log.	Cotangents Nat.	172°
0′	.12 187	9.08 589	1.7	.99 255	9.99 675	.02	.12 278	9.08 914	1.8	0.91 086	8.1448	60′
1	216	692		251	674	.08	308	9.09 019	1.7	0.90 981	248	59
2	245	795		248	672		338	123		877	054	58
3	274	897		244	670	.02	367	227		773	8.0860	57
4	302	999		240	669	.08	397	330		670	667	56
5	331	9.09 101		237	667	.02	426	434		566	476	55
6	360	202		233	666	.08	456	537		463	285	54
7	389	304		230	664	.02	485	640		360	095	53
8	418	405		226	663	.08	515	742		258	7.9906	52
9	447	506		222	661		544	845		155	718	51
10	.12 476	9.09 606	1.7	99 219	9.99 659	.02	.12 574	9.09 947	1.7	0.90 053	7.9530	50
11	504	707		215	658	.08	603	9.10 049		0.89 951	344	49
12	533	807		211	656	.02	633	150		850	.158	48
13	562	907		208	655	.08	662	252		748	7.8978	47
14	591	9.10 006		204	653		692	353		647	769	46
15	620	106		200	651	.02	722	454		546	606	45
16	649	205		197	650	.08	751	555		445	424	44
17	678	304	1.6	198	648	.02	781	656		344	248	43
18	706	402	1.7	189	647	.08	810	756		244	062	42
19	735	501	1.6	186	645		840	856		144	7.7882	41
20	.12 764	9.10 599	1.6	.99 182	9.99 643	.02	.12 809	9.10 956	1.7	0.89 044	7.7704	40
21	793	697		178	642	.08	899	9.11 056		0.88 944	525	39
22	822	795		175	640		929	155		845	848	38
23	851	893		171	638	.02	958	254		746	171	37
24	880	990		167	637	.03	989	353		647	7.6096	36
25	909	9.11 087		163	635		.13 017	452		548	821	35
26	937	184		160	633	.03	047	551	1.6	449	647	34
27	966	281		156	632	.03	076	649		351	473	33
28	995	377		152	630	.02	106	747		253	301	32
29	.13 024	474		148	629	.03	136	845		155	129	31
30	.13 053	9.11 570	1.6	.99 144	9.99 627	.08	.13 165	9.11 943	1.6	0.88 057	7.5958	30
31	081	666		141	625	.02	195	9.12 040		0.87 960	787	29
32	110	761		137	624	.08	224	138		862	618	28
33	139	857		133	622		254	235		765	449	27
34	168	952		129	620		284	332		668	281	26
35	197	9.12 047		125	618	.02	313	428		572	113	25
36	226	142		122	617	.08	343	525		475	7.4947	24
37	254	236		118	615		372	621		379	781	23
38	283	331		114	613	.02	402	717		283	615	22
39	312	425		110	612	.08	432	813		187	451	21
40	.13 341	9.12 519	1.6	.99 106	9.99 610	.08	.13 461	9.12 909	1.6	0.87 091	7.4287	20
41	370	612		102	608	.02	491	9.13 004		0.86 996	124	19
42	399	706		098	607	.08	521	099		901	7.3962	18
43	427	799		094	605		550	194		806	800	17
44	456	892		091	603		580	289		711	639	16
45	485	985		087	601	.02	609	384		616	479	15
46	514	9.13 078		083	600	.08	639	478		522	319	14
47	543	171	1.5	079	598		669	573		427	160	13
48	572	263		075	596	.02	698	667		333	002	12
49	600	355		071	595	.08	728	761		239	7.2844	11
50	.13 629	9.13 447	1.5	.99 067	9.99 593	.08	.13 758	9.13 854	1.6	0.86 146	7.2687	10
51	658	539		063	591		787	948		052	531	9
52	687	630		059	589	.02	817	9.14 041		0.85 959	375	8
53	716	722		055	588	.08	846	134		866	220	7
54	744	813		051	586		876	227		773	066	6
55	773	904		047	584		906	320	1.5	680	7.1912	5
56	802	994		043	582	.02	935	412		588	759	4
57	831	9.14 085		039	581	.08	965	504	1.6	496	607	3
58	860	175		035	579		995	597	1.5	403	455	2
59	889	266		031	577		.14 024	688		312	304	1
60	.13 917	9.14 356		.99 027	9.99 575		.14 054	9.14 780		0.85 220	7.1154	0

97°	Nat.	Log.	Dif.	Nat.	Log.	Dif.	Nat.	Log.	Dif.	Log.	Nat.	82°
	Cosines			Sines			Cotangents			Tangents		

8°	SINES Nat.	Log.	Dif.	COSINES Nat.	Log.	Dif.	TANGENTS Nat.	Log.	Dif.	COTANGENTS Log.	Nat.	171°
0′	.18 917	9.14 356	1.5	.99 027	9.99 575	.02	.14 054	9.14 780	1.5	0.85 220	7.1154	60′
1	946	445		023	574	.08	084	872		128	004	59
2	975	535		019	572		M8	963		037	7.0855	58
3	.14 004	624		015	570		143	9.15 054		0.84 946	706	57
4	033	714		011	568		173	145		855	558	56
5	061	803		006	566	.02	202	236		764	410	55
6	090	891		002	565	.08	232	327		673	264	54
7	119	980		.98 998	563		262	417		583	117	53
8	148	9.15 069		994	561		291	508		492	6.9972	52
9	177	157		990	559		321	598		402	827	51
10	.14 205	9.15 245	1.5	.98 986	9.99 557	.02	.14 351	9.15 688	1.5	0.84 312	6.9682	50
11	234	333		982	556	.08	381	777		223	536	49
12	263	421		978	554		410	867		133	395	48
13	292	508		978	552	.	440	956		044	252	47
14	320	596		969	550		470	9.16 046		0.83 954	110	46
15	349	683		965	548		499	135		865	6.8969	45
16	378	770		961	546	.02	529	224		776	828	44
17	407	857		957	545	.08	559	312		688	697	43
18	436	944	1.4	953	543		588	401		599	548	42
19	464	9.16 030		948	541		618	489		511	408	41
20	.14 493	9.16 116	1.5	.98 944	9.99 539	.08	.14 648	9.16 577	1.5	0.83 423	6.8260	40
21	522	203	1.4	940	537		678	665		335	181	39
22	551	289		936	535		707	753		247	6.7994	38
23	580	374		931	533	.02	737	841		159	656	37
24	608	460		927	532	.08	767	928		072	720	36
25	637	545		923	530		796	9.17 016		0.82 984	584	35
26	666	631		919	528		826	103		897	448	34
27	695	716		914	526		856	190		810	313	33
28	723	801		910	524		886	277	1.4	723	179	32
29	752	886		906	522		915	363	1.5	637	045	31
30	.14 781	9.16 970	1.4	.98 902	9.99 520	.08	.14 945	9.17 450	1.4	0.82 550	6.6912	30
31	810	9.17 055		897	518	.02	975	536		464	779	29
32	838	139		893	517	.08	.15 005	622		378	646	28
33	867	223		889	515		034	708		292	514	27
34	896	307		884	513		064	794		206	383	26
35	925	391		880	511		094	880		120	252	25
36	954	474		876	509		124	965		035	122	24
37	982	558		871	507		153	9.18 051		0.81 949	6.5992	23
38	.15 011	641		867	505		183	136		864	863	22
39	040	724		863	503		213	221		779	734	21
40	.15 069	9.17 807	1.4	.98 858	9.99 501	.08	.15 243	9.18 306	1.4	0.81 694	6.5606	20
41	097	890		854	499		272	391		609	478	19
42	126	973		849	497		302	475		525	350	18
43	155	9.18 055		845	495	.02	332	560		440	223	17
44	184	137		841	494	.08	362	644		356	097	16
45	212	220		836	492		391	728		272	6.4971	15
46	241	302		832	490		421	812		188	846	14
47	270	383		827	488		451	896		104	721	13
48	299	465		823	486		481	979		021	596	12
49	327	547		818	484		511	9.19 063		0.80 937	472	11
50	.15 356	9.18 628	1.4	.98 814	9.99 482	.08	.15 540	9.19 146	1.4	0.80 854	6.4348	10
51	385	709		809	480		570	229		771	225	9
52	414	790		805	478		600	312		688	103	8
53	442	871		800	476		630	395		605	6.3980	7
54	471	952		796	474		660	478		522	859	6
55	500	9.19 033	1.3	791	472		690	561		439	737	5
56	529	113		787	470		719	643		357	617	4
57	557	193		782	468		749	725		275	496	3
58	586	273		778	466		779	807		193	376	2
59	615	353		773	464		809	889		111	257	1
60	.15 643	9.19 433		.98 769	9.99 462		.15 838	9.19 971		0.80 029	6.3138	0
98°	Nat.	Log. COSINES.	Dif.	Nat.	Log. SINES.	Dif.	Nat.	Log. COTANGENTS.	Dif.	Log. TANGENTS.	Nat.	81°

9°	Sines.			Cosines.			Tangents.			Cotangents.		170°
	Nat.	Log.	Dif.	Nat.	Log.	Dif.	Nat.	Log.	Dif.	Log.	Nat.	
0'	.15 643	9.19 433	1.3	.98 769	9.99 462	.03	.15 838	9.19 971	1.4	0.80 029	6.3138	60'
1	'672	'513	'	764	460		868	9.20 053		0.79 947	019	59
2	701	592		760	458		898	134		866	6.2901	58
3	730	672		755	456		928	216		784	788	57
4	758	751		751	454		958	297		703	666	56
5	787	830		746	452		988	378		622	549	55
6	816	909		741	450		.16 017	459		541	432	54
7	845	988		737	448		047	540		460	316	53
8	873	9.20 067		732	446		077	621	1.3	379	200	52
9	902	145		728	444		107	701	1.4	299	085	51
10	.15 931	9.20 223	1.3	.98 728	9.99 442	.03	.16 137	9.20 782	1.3	0.79 218	6.1970	50
11	959	302		718	440		167	862		138	856	49
12	988	380		714	438		196	942		058	742	48
13	.16 017	458		709	436		226	9.21 022		0.78 978	628	47
14	046	535		704	434		256	102		898	515	46
15	074	613		700	432	.05	286	182		818	402	45
16	103	691		695	429	.03	316	261		739	290	44
17	132	768		690	427		346	341		659	178	43
18	160	845		686	425		376	420		580	006	42
19	189	922		681	423		405	499		501	6.0965	41
20	.16 218	9.20 999	1.3	.98 676	9.99 421	.03	.16 435	9.21 578	1.3	0.78 422	6.0844	40
21	246	9.21 076		671	419		465	657		343	734	39
22	275	153		667	417		495	736		264	624	38
23	304	229		662	415		525	814		186	514	37
24	333	306		657	413		555	893		107	405	36
25	361	382		652	411		585	971		029	296	35
26	390	458		648	409		615	9.22 049		0.77 951	188	34
27	419	534		643	407	.05	645	127		873	080	33
28	447	610		638	404	.03	674	205		795	5.9972	32
29	476	685		633	402		704	283		717	865	31
30	.16 505	9.21 761	1.3	.98 629	9.99 400	.03	.16 734	9.22 361	1.3	0.77 639	5.9758	30
31	533	836		624	398		764	438		562	651	29
32	562	912		619	396		794	516		484	545	28
33	591	987		614	394		824	593		407	439	27
34	620	9.22 062		609	392		854	670		330	333	26
35	648	137	1.2	604	390		884	747		253	228	25
36	677	211	1.3	600	388	.05	914	824		176	124	24
37	706	286		595	385	.03	944	901		099	019	23
38	734	361	1.2	590	383		974	977		023	5.9015	22
39	763	435		585	381		.17 004	9.23 054		0.76 946	811	21
40	.16 792	9.22 509	1.2	.98 580	9.99 370	.03	.17 033	9.23 130	1.3	0.76 870	5.8708	20
41	820	583		575	377		063	206		794	605	19
42	849	657		570	375	.05	093	283		717	502	18
43	878	731		565	372	.03	123	359		641	400	17
44	906	805		561	370		153	435		565	299	16
45	935	878		556	368		183	510		490	197	15
46	964	952		551	366		213	586		414	095	14
47	992	9.23 025		546	364		243	661		339	5.7994	13
48	.17 021	098		541	362	.05	273	737		263	894	12
49	050	171		536	359	.03	303	812		188	794	11
50	.17 078	9.23 244	1.2	.98 531	9.99 357	.03	.17 333	9.23 887	1.3	0.76 113	5.7694	10
51	107	317		526	355		363	962		038	594	9
52	136	390		521	353		393	9.24 037		0.75 963	495	8
53	164	462		516	351	.05	423	112	1.2	888	396	7
54	193	535		511	348	.08	453	186	1.3	814	297	6
55	222	607		506	346		483	261	1.2	739	199	5
56	250	679		501	344		513	335	1.3	665	101	4
57	279	752		496	342		543	410	1.2	590	004	3
58	308	823		491	340	.05	573	484		516	5.6906	2
59	336	895		486	337	.03	603	558		442	809	1
60	.17 865	9.23 967		.98 481	9.99 335		.17 633	9.24 632		0.75 368	5.6713	0

99°	Nat.	Log.	Dif.	Nat.	Log.	Dif.	Nat.	Log.	Dif.	Log.	Nat.	80°
		Cosines.			Sines.			Cotangents.			Tangents.	

10°	SINES.			COSINES.			TANGENTS.			COTANGENTS.		169°
	Nat.	Log.	Dif.	Nat.	Log.	Dif.	Nat.	Log.	Dif.	Log.	Nat.	
0'	.17 365	9.23 967	1.2	.98 481	9.99 335	.06	.17 633	9.24 632	1.2	0.75 368	5.6713	60'
1	393	9.24 039		476	333		663	706		294	617	59
2	422	110		471	331	.05	668	779		221	521	58
3	451	181		466	328	.03	723	853		147	425	57
4	479	253		461	326		758	926		074	329	56
5	508	324		455	324		788	9.25 000		000	234	55
6	537	395		450	322	.05	818	073		0.74 927	140	54
7	565	466		445	319	.03	848	146		854	045	53
8	594	536		440	317		673	219		781	5.5951	52
9	623	607		435	315		903	292		708	857	51
10	.17 651	9.24 677	1.2	.98 480	9.99 313	.05	.17 968	9.25 365	1.2	0.74 635	5.5764	50
11	680	748		425	310	.03	963	437		563	671	49
12	708	818		420	308		998	510		490	578	48
13	737	888		414	306		.18 028	582		418	485	47
14	766	958		409	304	.05	058	655		345	398	46
15	794	9.25 028		404	301	.03	088	727		273	301	45
16	823	098		399	299		118	799		201	209	44
17	852	168		394	297	.05	148	871		129	118	43
18	880	237		389	294	.03	178	943		057	026	42
19	909	307.		383	292		203	9.26 015		0.73 985	5.4936	41
20	.17 937	9.25 376	1.2	.98 378	9.99 290	.06	.18 238	9.26 086	1.2	0.73 914	5.4845	40
21	966	445		373	288	.05	263	158		842	755	39
22	995	514		368	285	.03	293	229		771	665	38
23	.18 023	583		362	283		323	301		699	575	37
24	052	652		357	281	.05	353	372		628	486	36
25	081	721		352	278	.03	384	443		557	897	35
26	109	790	1.1	347	276		414	514		486	808	34
27	138	858	1.2	341	274	.05	444	585		.415	219	33
28	166	927	1.1	336	271	.08	474	655		345	131	32
29	195	995		331	269		504	726		274	043	31
30	.18 224	9.26 063	1.1	.98 325	9.99 267	.05	.18 534	9.26 797	1.2	0.73 203	5.3955	30
31	252	131		320	264	.08	564	867		133	868	29
32	281	199		315	262		594	937		063	781	28
33	309	267		310	260	.05	624	9.27 008		0.72 992	694	27
34	338	335		304	·257	.08	654	078		922	607	26
35	367	403		299	255	.05	684	148		852	521	25
36	395	470		294	252	.08	714	218		782	435	24
37	424	538		288	250		745	288		712	349	23
38	452	605		283	248	.03	775	357		643	263	22
39	481	672		277	245	.08	805	427		573	178	21
40	.18 509	9.26 739	1.1	.98 272	9.99 243	.06	.18 835	9.27 496	1.2	0.72 504	5.3098	20
41	538	806		267	241	.05	865	566		434	008	19
42	567	873		261	238	.08	895	635		365	5.2924	18
43	595	940		256	236	.05	925	704		296	839	17
44	624	9.27 007		250	233	.03	955	773		227	755	16
45	652	073		245	231		986	842		158	672	15
46	681	140		240	229	.05	.19 016	911		089	588	14
47	710	206		234	226	.08	046	980		020	505	13
48	738	273		229	224	.05	076	9.28 049	1.1	0.71 951	422	12
49	767	339		228	221	.08	106	117	1.2	883	339	11
50	.18 795	9.27 405	1.1	.98 218	9.99 219	.06	.19 136	9.28 186	1.1	0.71 814	5.2257	10
51	824	471		212	217	.05	166	254	1.2	746	174	9
52	852	537		207	214	.08	197	323	1.1	677	092	8
53	881	602		201	212	.05	227	391		609	011	7
54	910	668		196	209	.08	257	459		541	5.1929	6
55	938	734		190	207	.05	287	527		473	848	5
56	967	799		185	204	.08	317	595		405	767	4
57	995	864		179	202		347	662		338	686	3
58	.19 024	930		174	200	.05	378	730		270	606	2
59	052	995		168	197	.05	408	798		202	526	1
60	.19 081	9.28 060		.98 163	9.99 195		.19 438	9.28 865		0.71 135	5.1446	0
100°	Nat.	Log.	Dif.	Nat.	Log.	Dif.	Nat.	Log.	Dif.	Log.	Nat.	79°
		COSINES.			SINES.			COTANGENTS.			TANGENTS.	

11°	Sines Nat.	Sines Log.	Dif.	Cosines Nat.	Cosines Log.	Dif.	Tangents Nat.	Tangents Log.	Dif.	Cotangents Log.	Cotangents Nat.	168°
0'	.19 081	9.28 060	1.1	.98 163	9.99 195	.05	.19 438	9.28 865	1.1	0.71 135	5.1446	60'
1	109	125		157	192	.03	468	933		067	366	59
2	138	190		152	190	.05	498	9.29 000		000	286	58
3	167	254		146	187	.03	529	067		0.70 933	207	57
4	105	319		140	185	.05	559	134		866	128	56
5	224	384		135	182	.03	589	201		799	049	55
6	252	448		129	180	.05	619	268		732	5.0970	54
7	281	512		124	177	.03	649 *	335		665	692	53
8	309	577		118	175	.05	680	402		598	614	52
9	338	641		112	172	.03	710	468		532	736	51
10	.19 366	9.28 705	1.1	.98 107	9.99 170	.05	.19 740	9.29 535	1.1	0.70 465	5.0658	50
11	395	769		101	167	.03	770	601		399	561	49
12	423	833		096	165	.05	801	668		332	504	48
13	452	896		090	162	.03	831	734		266	427	47
14	481	960		084	160	.05	861	800		200	350	46
15	509	9.29 024		079	157	.03	891	866		134	273	45
16	538	087		073	155	.05	921	932		068	197	44
17	566	150		067	152	.03	952	998		002	121	43
18	595	214		061	150	.05	982	9.30 064		0.69 936	045	42
19	623	277		056	147	.03	.20 012	130		870	4.9969	41
20	.19 652	9.29 340	1.1	.98 050	9.99 145	.05	.20 042	9.30 195	1.1	0.69 805	4.9694	40
21	680	403		044	142	.03	073	261		739	819	39
22	709	466		080	140	.05	103	326		674	744	38
23	737	529	1.0	083	137	.03	133	391		609	669	37
24	766	591	1.1	027	135	.05	164	457		543	594	36
25	794	654	1.0	021	132	.03	194	522		478	520	35
26	823	716	1.1	016	130	.05	224	587		413	446	34
27	851	779	1.0	010	127		254	652		348	372	33
28	880	841		004	124	.03	285	717		283	298	32
29	908	903	1.1	.97 999	122	.05	315	782		218	225	31
30	.19 937	9.29 966	1.0	.97 993	9.99 119	.03	.20 345	9.30 846	1.1	0.69 154	4.9152	30
31	965	9.30 028		987	117	.05	376	911		089	078	29
32	994	090		981	114	.03	406	975		025	006	28
33	.20 022	151		975	112	.05	436	9.31 040		0.68 960	4.8933	27
34	051	213		969	109		466	104		896	860	26
35	079	275		963	106	.03	497	168		832	788	25
36	108	336		958	104	.05	527	233		767	716	24
37	136	398		952	101	.03	557	297		703	644	23
38	165	459		946	099	.05	588	361		639	573	22
39	193	521		940	096		618	425		575	501	21
40	.20 222	9.30 582	1.0	.97 934	9.99 093	.03	.20 648	9.31 489	1.1	0.68 511	4.8430	20
41	250	643		928	091	.05	679	552		448	359	19
42	279	704		922	088	.03	709	616		384	288	18
43	307	765		916	086	.05	739	679		321	216	17
44	336	826		910	083		770	743		257	147	16
45	364	887		905	080	.03	800	806		194	077	15
46	393	947		899	078	.05	830	870		130	007	14
47	421	9.31 008		893	075		861	933		067	4.7937	13
48	450	068		887	072	.03	891	996		004	867	12
49	478	129		881	070	.05	921	9.32 059		0.67 941	798	11
50	.20 507	9.31 189	1.0	.97 875	9.99 067	.05	.20 952	9.32 122	1.1	0.67 878	4.7729	10
51	535	250		869	064	.03	982	185		815	659	9
52	563	310		863	062	.05	.21 013	248		752	591	8
53	592	370		857	059		043	311	1.0	689	522	7
54	620	430		851	056	.03	073	373	1.1	627	453	6
55	649	490	.98	845	054	.05	104	436	1.0	564	385	5
56	677	549	1.0	839	051		134	498	1.1	502	317	4
57	706	609		833	048	.03	164	561	1.0	439	249	3
58	734	669	.98	827	046	.05	195	623		377	181	2
59	763	728	1.0	821	043		225	'685		315	114	1
60	.20 791	9.31 788		.97 815	9.99 040		.21 256	9.32 747		0.67 253	4.7046	0
101°	Nat.	Log.	Dif.	Nat.	Log.	Dif.	Nat.	Log.	Dif.	Log.	Nat.	78°
		Cosines.			Sines.			Cotangents.			Tangents.	

12°	SINES.			COSINES.			TANGENTS.			COTANGENTS.		167°
	Nat.	Log.	Dif.	Nat.	Log.	Dif.	Nat.	Log.	Dif.	Log.	Nat.	
0'	.20 791	9.31 788	.9S	.97 515	9.99 040	.08	.21 256	9.32 747	1.1	0.67 253	4.7046	60'
1	820	847	1.0	809	038	.05	286	810	1.0	190	4.6979	59
2	848	907	.9S	808	035		316	872		128	912	58
3	877	966		797	032	.08	847	933		067	845	57
4	905	9.32 025		791	030	.05	377	995		005	779	56
5	933	084		784	027		408	9.33 057		0.66 943	712	55
6	962	143		778	024	.08	438	119		881	646	54
7	990	202		772	022	.05	469	180		820	580	53
8	.21 019	261	.97	766	019		499	242		758	514	52
9	047	319	.9S	760	016		529	303		697	448	51
10	.21 076	9.32 378	.9S	.97 754	9.99 013	.08	.21 560	9.33 365	1.0	0.66 635	4.6882	50
11	104	437	.97	748	011	.05	590	426		574	817	49
12	132	495		742	008		621	487		513	252	48
13	161	553	.9S	735	005		651	548		452	187	47
14	189	612	.97	729	002	.08	682	609		391	122	46
15	218	670		723	000	.05	712	670		330	057	45
16	246	728		717	9.98 997		743	731		269	4.5993	44
17	275	786		711	994		773	792		208	928	43
18	303	844		705	991	.08	804	853		147	864	42
19	331	902		698	989	.05	834	913		087	800	41
20	.21 360	9.32 960	.97	.97 692	9.98 986	.05	.21 864	9.33 974	1.0	0.66 026	4.5736	40
21	388	9.33 018	.95	686	983		895	9.34 034		0.65 966	673	39
22	417	075	.97	680	980	.08	925	095		905	609	38
23	445	133	.95	673	978	.05	956	155		845	546	37
24	474	190	.97	667	975		986	215		785	483	36
25	502	248	.95	661	972		.22 017	276		724	420	35
26	530	305		655	969	.08	047	336		664	357	34
27	559	362	.97	649	967	.05	078	396		604	294	33
28	587	420	.95	642	964		108	456		544	232	32
29	616	477		636	961		139	516		484	169	31
30	.21 644	9.33 534	.95	.97 630	9.98 958	.05	.22 169	9.34 576	.9S	0.65 424	4.5107	30
31	672	591	.93	623	955	.08	200	635	1.0	365	045	29
32	701	647	.95	617	953	.05	231	695		305	4.4988	28
33	729	704		611	950		261	755	.98	245	922	27
34	758	761		604	947		292	814	1.0	186	860	26
35	786	818	.98	598	944		322	874	.98	126	799	25
36	814	874	.95	592	941		353	933		067	737	24
37	843	931	.98	585	938	.08	383	992		008	676	23
38	871	987		579	936	.05	414	9.35 051	1.0	0.64 949	615	22
39	899	9.34 043	.95	573	933		444	111	.98	889	555	21
40	.21 928	9.34 100	.98	.97 566	9.98 930	.05	.22 475	9.35 170	.98	0.64 830	4.4494	20
41	956	156		560	927		505	229		771	434	19
42	985	212		553	924		536	288		712	373	18
43	.22 013	268		547	921	.08	567	347	.97	653	313	17
44	041	324		541	919	.05	597	405	.98	595	253	16
45	070	380		534	916		628	464		536	194	15
46	098	436	.92	528	913		658	523	.97	477	134	14
47	126	491	.93	521	910		689	581	.98	419	075	13
48	155	547	.92	515	907		719	640	.97	360	015	12
49	183	602	.99	509	904		750	698	.98	302	4.3956	11
50	.22 212	9.34 658	.92	.97 502	9.98 901	.05	.22 781	9.35 757	.97	0.64 243	4.3897	10
51	240	713	.93	496	898		811	815		185	838	9
52	268	769	.92	489	896	.05	842	873		127	779	8
53	297	824		483	893		872	931		069	721	7
54	325	879		476	890		903	989		011	662	6
55	353	934		470	887		934	9.36 047		0.63 953	604	5
56	382	989		463	884		964	105		895	546	4
57	410	9.35 044		457	881		995	163		837	488	3
58	438	099		450	878		.23 026	221		779	480	2
59	467	154		444	875		056	279	.95	721	872	1
60	.22 495	9.35 209		.97 487	9.98 872		.23 087	9.36 336		0.63 664	4.3815	0

13°	SINES.			COSINES.			TANGENTS.			COTANGENTS.		166°
	Nat.	Log.	Dif.	Nat.	Log.	Dif.	Nat.	Log.	Dif.	Log.	Nat.	
0'	.22 495	9.35 209	.90	.97 487	9.98 872	.05	.23 057	9.36 336	.97	0.63 664	4.3315	60'
1	523	263	.92	430	869	.03	117	394		606	257	59
2	552	318		424	867	.05	148	452	.95	548	200	58
3	580	373	.90	417	864		179	509		491	143	57
4	608	427		411	861		209	566	.97	434	056	56
5	637	481	.92	404	858		240	624	.95	376	029	55
6	665	536	.90	398	855		271	681		319	4.2972	54
7	693	590		391	852		301	738		262	916	53
8	722	644		384	849		332	795		205	859	52
9	750	698		378	846		363	852		148	803	51
10	.22 778	9.35 752	.90	.97 371	9.98 843	.05	.23 393	9.36 909	.95	0.63 091	4.2747	50
11	807	806		365	840		424	966		034	691	49
12	835	860		358	837		455	9.37 023		0.62 977	635	48
13	863	914		351	834		485	080		920	580	47
14	892	968		345	831		516	137	.93	863	524	46
15	920	9.36 022	.88	338	828		547	193	.95	807	468	45
16	948	075	.90	331	825		578	250	.93	750	413	44
17	977	129	.88	325	822		608	306	.05	694	358	43
18	.23 005	182	.90	318	819		639	363	.08	637	303	42
19	033	236	.88	311	816		670	419	.05	581	248	41
20	.23 062	9.36 289	.88	.97 304	9.98 813	.05	.23 700	9.37 476	.93	0.62 524	4.2193	40
21	090	342		298	810		731	532		468	139	39
22	118	395	.90	291	807		702	588		412	084	38
23	146	449	.88	284	804		793	644		356	080	37
24	175	502		278	801		823	700		300	4.1976	36
25	203	555		271	798		854	756		244	922	35
26	231	608	.87	264	795		885	812		188	868	34
27	260	660	.88	257	792		916	868		132	814	33
28	288	713		251	789		946	924		076	760	32
29	316	766		244	786		977	980	.92	020	706	31
30	.23 345	9.36 819	.87	.97 237	9.98 783	.05	.24 008	9.38 035	.93	0.61 965	4.1653	30
31	373	871	.88	230	780		089	091		909	600	29
32	401	924	.87	223	777		069	147	.92	853	547	28
33	429	976		217	774		100	202		798	493	27
34	458	9.37 028	.88	210	771		131	257	.93	743	441	26
35	486	081	.87	203	768		162	313	.92	687	388	25
36	514	133		196	765		193	368		632	335	24
37	542	185		189	762		223	423	.93	577	282	23
38	571	237		182	759		254	479	.92	521	230	22
39	599	289		176	756		265	534		466	178	21
40	.23 627	9.37 341	.87	.97 169	9.98 753	.05	.24 316	9.38 589	.92	0.61 411	4.1126	20
41	656	393		162	750	.07	347	644		356	074	19
42	684	445		155	746	.05	377	699		301	022	18
43	712	497		149	743		408	754	.90	246	4.0970	17
44	740	549	.85	141	740		439	808	.92	192	918	16
45	769	600	.87	134	737		470	863		137	667	15
46	797	652	.85	127	734		501	918	.90	082	815	14
47	825	703	.87	120	731		532	972	.92	028	764	13
48	853	755	.85	113	728		562	9.39 027		0.60 973	713	12
49	882	806	.87	106	725		593	082	.90	918	662	11
50	.23 910	9.37 858	.85	.97 100	9.98 722	.05	.24 624	9.39 136	.90	0.60 864	·4.0011	10
51	938	909		093	719	.07	655	190	.02	810	560	9
52	966	960		086	715	.05	686	245	.90	755	509	8
53	995	9.38 011		079	712		717	299		701	459	7
54	.24 023	062		072	709		747	353		647	408	6
55	051	113		065	706		778	407		593	358	5
56	079	164		058	703		809	461		539	308	4
57	108	215		051	700		840	515		485	257	3
58	136	266		044	697		871	569		431	207	2
59	164	317		037	694	.07	902	623		377	158	1
60	.24 192	9.38 368		.97 030	9.98 690		.24 933	9.39 677		0.60 323	4.0108	0
103°	Nat.	Log.	Dif.	Nat.	Log.	Dif.	Nat.	Log.	Dif.	Log.	Nat.	76°
		COSINES.			SINES.			COTANGENTS.			TANGENTS.	

14°	SINES.			COSINES.			TANGENTS.			COTANGENTS.		165°
	Nat.	Log.	Dif.	Nat.	Log.	Dif.	Nat.	Log.	Dif.	Log.	Nat.	
0'	.24192	9.38368	.83	.97080	9.98690	.05	.24933	9.39677	.90	0.60323	4.0108	60'
1	220	418	.85	023	687		964	731		269	058	59
2	249	469	.83	015	684		995	785	.88	215	009	58
3	277	519	.85	008	681		.25026	838	.90	162	3.9959	57
4	305	570	.83	001	678		056	892	.88	108	910	56
5	333	620		.96994	675	.07	087	945	.90	055	861	55
6	362	670	.85	987	671	.05	118	999	.88	001	812	54
7	390	721	.83	980	668		149	9.40052	.90	0.59948	763	53
8	418	771		973	665		180	106	.88	894	714	52
9	446	821		966	662		211	159		841	665	51
10	.24474	9.38871	.83	.96959	9.98659	.05	.25242	9.40212	.90	0.59788	8.9617	50
11	503	921		952	656	.07	273	266	.88	734	568	49
12	531	971		945	652	.05	304	319		681	520	48
13	559	9.39021		937	649		335	372		628	471	47
14	587	071		930	646		366	425		575	423	46
15	615	121	.82	923	643		397	478		522	375	45
16	644	170	.83	916	640	.07	428	531		469	327	44
17	672	220		909	636	.05	459	584	.87	416	279	43
18	700	270	.82	902	633		490	636	.88	364	232	42
19	728	319	.83	894	630		521	689		311	184	41
20	.24756	9.39369	.82	.96887	9.98627	.07	.25552	9.40742	.88	0.59258	3.9136	40
21	784	418		860	623	.05	583	795	.87	205	069	39
22	813	467	.83	873	620		614	847	.88	153	042	38
23	841	517	.82	866	617		645	900	.87	100	3.8995	37
24	869	566		858	614	.07	676	952	.88	048	947	36
25	897	615		851	610	.05	707	9.41005	.87	0.58995	900	35
26	925	664		844	607		738	057		943	854	34
27	954	713		837	604		769	109		891	807	33
28	982	762		829	601	.07	800	161	.86	839	760	32
29	.25010	811		822	597	.05	831	214	.87	786	714	31
30	.25038	9.39860	.82	.96815	9.98594	.05	.25862	9.41266	.87	0.58734	3.8667	30
31	066	909		807	591		893	318		682	621	29
32	094	958	.80	800	588	.07	924	370		630	575	28
33	122	9.40006	.82	793	584	.05	955	422		578	528	27
34	151	055	.80	786	581		986	474		526	482	26
35	179	103	.82	778	578	.07	.26017	526		474	486	25
36	207	152	.80	771	574	.05	048	578	.85	422	391	24
37	235	200	.82	764	571		079	629	.87	371	345	23
38	263	249	.80	756	568		110	681		319	299	22
39	291	297	.82	749	565	.07	141	733	.85	267	254	21
40	.25320	9.40346	.80	.96742	9.98561	.05	.26172	9.41784	.87	0.58216	3.8208	20
41	348	394		734	558		203	836	.85	164	163	19
42	376	442		727	555	.07	235	887	.87	113	118	18
43	404	490		719	551	.05	266	939	.85	061	073	17
44	432	538		712	548		297	990		010	028	16
45	460	586		705	545	.07	328	9.42041	.87	0.57959	8.7958	15
46	488	634		697	541	.05	359	093	.85	907	938	14
47	516	682		690	538		390	144		856	893	13
48	545	730		682	535	.07	421	195		805	848	12
49	573	778	.79	675	531	.05	452	246		754	804	11
50	.25601	9.40825	.80	.96667	9.98528	.05	.26483	9.42297	.85	0.57703	3.7760	10
51	629	873		660	525	.07	515	348		652	715	9
52	657	921	.79	653	521	.05	546	399		601	671	8
53	685	968	.80	645	518		577	450		550	627	7
54	713	9.41016	.79	638	515	.07	608	501		499	588	6
55	741	063	.80	630	511	.05	639	552		448	539	5
56	769	111	.78	623	508		670	603	.83	397	495	4
57	798	158		615	505	.07	701	653	.85	347	451	3
58	826	205		608	501	.05	732	704		296	408	2
59	854	252	.80	600	498	.07	764	755	.83	245	364	1
60	.25882	9.41300		.96593	9.98494		.26795	9.42805		0.57195	3.7321	0

104°	Nat.	Log.	Dif.	Nat.	Log.	Dif.	Nat.	Log.	Dif.	Log.	Nat.	75°
		COSINES.			SINES.			COTANGENTS.			TANGENTS.	

15°	Sines.			Cosines.			Tangents.			Cotangents.		164°
	Nat.	Log.	Dif.	Nat.	Log.	Dif.	Nat.	Log.	Dif.	Log.	Nat.	
0′	.25 882	9.41 300	.78	.96 598	9.98 494	.05	.26 795	9.42 805	.85	0.57 195	3.7321	60′
1	910	347		585	491		826	856	.83	144	277	59
2	938	394		578	488	.07	857	906	.85	094	234	58
3	966	441		570	484	.05	888	957	.83	043	191	57
4	994	488		562	481	.07	920	9.43 007		0.56 993	148	56
5	.26 022	535		· 555	477	.05	951	057	.85	943	105	55
6	050	582	.77	547	474		982	108	.83	892	062	54
7	079	628	.78	540	471	.07	.27 013	158		842	019	53
8	107	675		532	467	.05	044	208		792	3.6976	52
9	135	722	.77	524	464	.07	076	258		742	933	51
10	.26 163	9.41 768	.78	.96 517	9.98 460	.05	.27 107	9.43 308	.83	0.56 692	3.6591	50
11	191	815		. 509	457	.07	138	358		642	848	49
12	219	861	.78	502	453	.05	169	408		592	806	48
13	247	908	.77	494	450		201	458		542	764	47
14	275	954	.78	486	447	.07	232	508		492	722	46
15	303	9.42 001	.77	479	443	.05	263	558	.82	442	680	45
16	331	047		471	440	.07	294	607	.83	393	638	44
17	359	093	.78	463	436	.05	326	657		343	596	43
18	387	140	.77	456	433	.07	357	707	.82	293	554	42
19	415	186		448	429	.05	388	756	.83	244	512	41
20	.26 443	9.42 232	.77	.96 440	9.98 426	.07	.27 419	9.43 806	.82	0.56 194	3.6470	40
21	471	278		433	422	.05	451	855	.83	145	429	39
22	500	324		425	419	.07	482	905	.82	095	387	38
23	528	370		417	415	.05	513	954	.83	046	346	37
24	556	416	.75	410	412		545	9.44 004	.82	0.55 996	305	36
25	584	461	.77	402	409	.07	576	053		947	264	35
26	612	507		394	405	.05	607	102		898	222	34
27	640	553·		386	402	.07	638	151	.58	849	181	33
28	668	599	.75	379	398	.05	670	201	.82	799	140	32
29	696	644	.77	371	395	.07	701	250		750	100	31
30	.26 724	9.42 690	.75	.96 363	9.98 391	.05	.27 732	9.44 299	.82	0.55 701	3.6059	30
31	752	735	.77	355	388	.07	764	348		652	018	29
32	780	781	.78	347	384	.05	795	397		603	3.5978	28
33	808	826	.77	340	381	.07	826	446		554	937	27
34	836	872	.75	332	377		859	495		505	697	26
35	864	917		324	373	.05	880	544	.60	456	856	25
36	892	962	.77	316	370	.07	921	592	.82 ·	408	816	24
37	920	9.43 008	.75	309	366	.05	952	641		359	776	23
38	948	053		301	363	.07	. 983	690	.60	310	736	22
39	976	098		293	359	.05	.28 015	738	.62	262	696	21
40	.27 004	9.43 143	.75	.96 285	9.98 356	.07	.28 046	9.44 787	.82	0.55 213	3.5656	20
41	032	188		277	352	.05	077	836	.80	164	616	19
42	060	233		269	349	.07	109	884	.82	116	576	18
43	088	278		261	345	.05	140	933	.80	067	536	17
44	116	323	.78	253	342	.07	. 172	981		019	497	16
45	144	367	.75	246	338		203	9.45 029	.82	0.54 971	457	15
46	172	412		238	334	.05	234	078	.60	922	418	14
47	200	457		230	331	.07	266	126		874	379	13
48	228	502	.78	222	327	.05	297	174		826	339	12
49	256	546	.75	214	324	.07	329	222	.62	778	300	11
50	.27 284	9.43 591	· .78	.96 206	9.98 320	.05	.28 360	9.45 271	.60	0.54 729	3.5261	10
51	312	635	.75	198	317	.07	391	319		681	222	9
52	340	680	.78	190	313		423 ·	367		633	183	8
53	368	724	.75	182	309	.05	454	415		585	144	7
54	396	769	· .78	174	306	.07	486	463		537	105	6
55	424	813		166	302	.05	517	511		489	067	5
56	452	857		158	299	.07	549	559	.76	441	028	4
57	480	901	.75	150	295		580	606	.80	394	3.4989	3
58	508	946	.78	142˙	291	.05	612	654		346	951	2
59	536	990		134	288	.07	643	702		298	912	1
60	.27 564	9.44 034		.96 126	9.98 284		.28 675	9.45 750		0.54 250	3.4874	0

105°	Nat.	Log.	Dif.	Nat.	Log.	Dif.	Nat.	Log.	Dif.	Log.	Nat.	74°
		Cosines.			Sines.			Cotangents.			Tangents.	

16°	Sines.			Cosines.			Tangents.			Cotangents.		163°
	Nat.	Log.	Dif.	Nat.	Log.	Dif.	Nat.	Log.	Dif.	Log.	Nat.	
0′	.27 564	9.44 034	.73	.96 126	9.98 284	.05	.28 675	9.45 750	.78	0.54 250	3.4674	60′
1	592	078		118	281	.07	706	797	.80	203	636	59
2	620	122		110	277		738	845	.78	155	798	58
3	648	166		102	273	.05	769	892	.80	108	760	57
4	676	210	.72	094	270	.07	801	940	.78	060	722	56
5	704	253	.73	086	266		832	987	.80	013	684	55
6	731	297		078	262	.05	864	9.46 035	.78	0.53 965	640	54
7	759	341		070	259	.07	895	082	.80	918	608	53
8	787	385	.72	062	255		927	130	.78	870	570	52
9	815	428	.73	054	251	.05	958	177		823	533	51
10	.27 843	9.44 472	.73	.96 046	9.98 248	.07	.28 990	9.46 224	.78	0.53 776	3.4495	50
11	871	516	.72	037	244		.29 021	271	.80	729	458	49
12	899	559		029	240	.05	053	319	.78	681	420	48
13	927	602	.73	021	237	.07	084	366		634	383	47
14	955	646	.72	013	233		116	413		587	346	46
15	983	689	.73	005	229	.05	147	460		540	308	45
16	.28 011	733	.72	.95 997	226	.07	179	507		493	271	44
17	039	776		989	222		210	554		446	234	43
18	067	819		981	218	.05	242	601		399	197	42
19	095	862		972	215	.07	274	648	.77	352	160	41
20	.28 123	9.44 905	.72	.95 964	9.98 211	.07	.29 305	9.46 694	.78	0.53 306	3.4124	40
21	150	948	.73	956	207	.05	337	741		259	087	39
22	178	992	.72	948	204	.07	368	788		212	050	38
23	206	9.45 035	.70	940	200		400	835	.77	165	014	37
24	234	077	.72	931	196		432	881	.78	119	3.3977	36
25	262	120		923	192	.05	463	928		072	941	35
26	290	163		915	189	.07	495	975	.77	025	904	34
27	319	206		907	185		526	9.47 021	.78	0.52 979	868	33
28	346	249		898	181		558	068	.77	932	832	32
29	374	292	.70	890	177	.05	590	114		886	796	31
30	.28 402	9.45 334	.72	.95 882	9.98 174	.07	.29 621	9.47 160	.78	0.52 840	3.3759	30
31	429	377	.70	874	170		653	207	.77	793	723	29
32	457	419	.72	865	166		685	253		747	687	28
33	485	462	.70	857	162	.05	716	299	.78	701	652	27
34	513	504	.72	849	159	.07	748	346	.77	654	616	26
35	541	547	.70	841	155		780	392		608	580	25
36	569	589	.72	832	151		811	438		562	544	24
37	597	632	.70	824	147	.05	843	484		516	509	23
38	625	674		816	144	.07	875	530		470	473	22
39	652	716		807	140		906	576		424	437	21
40	.28 680	9.45 758	.72	.95 799	9.98 136	.07	.29 938	9.47 622	.77	0.52 378	3.3402	20
41	708	801	.70	791	132	.05	970	668		332	367	19
42	736	843		782	129	.07	.30 001	714		286	332	18
43	764	885		774	125		033	760		240	297	17
44	792	927		766	121		065	806		194	261	16
45	820	969		757	117		097	852	.75	148	226	15
46	847	9.46 011		749	113	.05	129	897	.77	103	191	14
47	875	053		740	110	.07	160	943		057	156	13
48	903	095	.69	732	106		192	989		011	122	12
49	931	136	.70	724	102		224	9.48 035	.75	0.51 965	087	11
50	.28 959	9.46 178	.70	.95 715	9.98 098	.07	.30 255	9.48 080	.77	0.51 920	3.3052	10
51	987	220		707	094		287	126	.75	874	017	9
52	.29 015	262	.68	698	090	.05	319	171	.77	829	3.2983	8
53	042	303	.70	690	087	.07	351	217	.75	783	948	7
54	070	345	.68	681	083		382	262		738	914	6
55	098	386	.70	673	079		414	307	.77	693	879	5
56	126	428	.68	664	075		446	353	.75	647	845	4
57	154	469	.70	656	071		478	398		602	811	3
58	182	511	.68	647	067		500	443	.77	557	777	2
59	209	552	.70	639	063	.05	541	489	.75	511	743	1
60	.29 237	9.46 594		.95 630	9.98 060		.30 573	9.48 534		0.51 466	3.2709	0

106°	Nat.	Log.	Dif.	Nat.	Log.	Dif.	Nat.	Log.	Dif.	Log.	Nat.	73°
		Cosines.			Sines.			Cotangents.			Tangents.	

17°	SINES.			COSINES.			TANGENTS.			COTANGENTS.		162°
	Nat.	Log.	Dif.	Nat.	Log.	Dif.	Nat.	Log.	Dif.	Log.	Nat.	
0'	.29 237	9.46 594	.68	.95 630	9.98 060	.07	.30 573	9.48 534	.75	0.51 466	3.2709	60'
1	265	635		622	056		605	579		421	675	59
2	298	676		613	052		637	624		376	641	58
3	321	717		605	048		669	669		331	607	57
4	348	758	.70	596	044		700	714		286	573	56
5	376	800	.68	588	040		732	759		241	539	55
6	404	841		579	036		764	804		196	506	54
7	432	882		571	032	.05	796	849		151	472	53
8	460	923		562	029	.07	828	894		106	438	52
9	487	964		554	025		800	939		061	405	51
10	.29 515	9.47 005	.67	.95 545	9.98 021	.07	.30 891	9.48 984	.75	0.51 016	3.2371	50
11	543	045	.68	536	017		923	9.49 029	.73	0.50 971	338	49
12	571	086		528	013		955	073	.75	927	305	48
13	599	127		519	009		987	118		882	272	47
14	626	168		511	005		.31 019	163	.73	837	238	46
15	654	209	.67	502	001		. 051	207	.75	793	205	45
16	682	249	.68	493	9.97 997		083	. 252	.73	748	172	44
17	710	290	.67	485	993		115	296	.75	704	139	43
18	737	330	.68	476	989	.05	147	341	.73	659	106	42
19	765	371	.67	467	986	.07	178	385	.75	615	073	41
20	.29 793	9.47 411	.68	.95 459	9.97 982	.07	.31 210	9.49 430	.73	0.50 570	3.2041	40
21	821	452	.67	450	978		242	474	.75	526	008	39
22	849	492	.68	441	974		274	519	.73	481	3.1975	38
23	876	533	.67	433	970		306	563		437	943	37
24	904	573		424	966		838	607	.75	393	910	36
25	932	613	.68	415	962		370	652	.73	348	878	35
26	960	654	.67	407	958		402	696		304	845	34
27	987	694		398	954		434	740		260	813	33
28	.30 015	734		389	950		466	784		216	780	32
29	043	774		380	946		498	828		172	748	31
30	.30 071	9.47 814	.67	.95 372	9.97 942	.07	.31 530	9.49 872	.73	0.50 128	3.1716	30
31	098	854		363	938		562	916		084	664	29
32	126	894		354	934		594	960		040	652	28
33	154	934		345	930		626	9.50 004		0.49 996	620	27
34	182	974		337	926		658	048		952	588	26
35	209	9.48 014		328	922		690	092		908	556	25
36	237	054		319	918		722	136		864	524	24
37	265	094	.65	310	914		754	180	.72	820	492	23
38	292	133	.07	301	910		786	223	.73	777	460	22
39	320	173	.	298	906		818	267		733	429	21
40	.30 348	9.48 213	.65	.95 284	9.97 902	.07	.31 850	9.50 311	.73	0.49 689	3.1397	20
41	376	252	.67	275	898		882	355	.72	645	366	19
42	408	292		266	894		914	398	.73	602	334	18
43	431	332	.65	257	890		946	442	.72	558	303	17
44	459	371	.67	248	886		978	485	.73	515	271	16
45	486	411	.65	240	882		.32 010	529	.72	471	240	15
46	514	450	.07	231	878		042	572	.73	428	209	14
47	542	490	.65	222	874		074	616	.72	384	178	13
48	570	529		213	870		106	659	.73	341	146	12
49	597	568		204	866	.06	139	703	.72	297	115	11
50	.30 625	9.48 607	.67	.95 195	9.97 861	.07	.32 171	9.50 746	.72	0.49 254	3.1064	10
51	653	647	.65	186	857		203	789	.73	211	053	9
52	680	686		177	853		235	833	.72	167	022	8
53	708	725		168	849		267	876		124	3.0991	7
54	736	764		159	845		299	919		081	961	6
55	763	803		150	841		331	962		038	930	5
56	791	842		142	837		363	9.51 005		0.48 995	899	4
57	819	881		133	833		396	048	.73	952	868	3
58	846	920		124	829		428	092	.72	908	838	2
59	874	959		115	825		460	135		865	807	1
60	.30 902	9.48 998		.95 106	9.97 821		.32 492	9.51 178		0.48 822	3.0777	0
107°	Nat.	Log.	Dif.	Nat.	Log.	Dif.	Nat.	Log.	Dif.	Log.	Nat.	72°
		COSINES.			SINES.			COTANGENTS.		TANGENTS.		

TRIGONOMETRIC FUNCTIONS.

18°	Sines.			Cosines.			Tangents.			Cotangents.		161°
	Nat.	Log.	Dif.	Nat.	Log.	Dif.	Nat.	Log.	Dif.	Log.	Nat.	
0′	.30 902	9.48 998	.65	.95 106	9.97 821	.07	.32 492	9.51 178	.72	0.48 822	3.0777	60′
1	929	9.49 037		097	817	.06	524	221		779	746	59
2	957	076		088	812	.07	556	264	.70	736	716	58
3	985	115	.63	079	808		588	306	.72	694	686	57
4	.31 012	153	.65	070	804		621	349		651	655	56
5	040	192		061	800		653	392		608	625	55
6	068	231	.63	052	796		685	435		565	595	54
7	095	269	.65	043	792		717	478	.70	522	565	53
8	123	308		033	788		749	520	.72	480	535	52
9	151	347	.63	024	784	.08	782	563		437	505	51
10	.31 178	9.49 385	.65	.95 015	9.97 779	.07	.32 814	9.51 606	.70	0.48 394	3.0475	50
11	206	424	.63	006	775		846	648	.72	352	445	49
12	233	462		.94 997	771		878	691		309	415	48
13	261	500	.65	988	767		911	734	.70	266	385	47
14	289	539	.63	979	763		943	776	.72	224	356	46
15	316	577		970	759	.08	975	819	.70	181	326	45
16	344	615	.65	961	754	.07	.33 007	861		139	296	44
17	372	654	.63	952	750		040	903	.72	097	267	43
18	399	692		943	746		072	946	.70	054	237	42
19	427	730		933	742		104	988	.72	012	208	41
20	.31 454	9.49 768	.63	.94 924	9.97 738	.07	.33 136	9.52 031	.70	0.47 969	3.0178	40
21	482	806		915	734	.09	169	073		927	149	39
22	510	844		906	729	.07	201	115		885	120	38
23	537	882		897	725		233	157	.72	843	090	37
24	565	920		888	721		266	200	.70	800	061	36
25	598	958		878	717		295	242		758	032	35
26	620	996		869	713	.08	330	284		716	003	34
27	648	9.50 034		860	708	.07	363	326		674	2.9974	33
28	675	072		851	704		395	368		632	945	32
29	703	110		842	700		427	410		590	916	31
30	.31 730	9.50 148	.62	.94 832	9.97 696	.08	.33 460	9.52 452	.70	0.47 548	2.9887	30
31	758	185	.63	828	691	.07	492	494		506	858	29
32	786	223		814	687		524	536		464	829	28
33	813	261	.62	805	683		557	578		422	800	27
34	841	298	.63	795	679	.08	589	620	.68	380	772	26
35	868	336		786	674	.07	621	661	.70	339	743	25
36	896	374	.62	777	670		654	703		297	714	24
37	923	411	.68	768	666		686	745		255	686	23
38	951	449	.62	758	662	.08	718	787		213	657	22
39	979	486		749	657	.07	751	829	.68	171	629	21
40	.32 006	9.50 523	.68	.94 740	9.97 653	.07	.33 783	9.52 870	.70	0.47 130	2.9000	20
41	034	561	.02	730	649		816	912	.68	088	572	19
42	061	598		721	645	.08	848	953	.70	047	544	18
43	089	635	.63	712	640	.07	881	995		005	515	17
44	116	673	.62	702	636		913	9.53 037	.68	0.46 963	487	16
45	144	710		693	632		945	078	.70	922	458	15
46	171	747		684	628	.06	978	120	.68	880	431	14
47	199	784		674	623	.07	.34 010	161	.	839	408	13
48	227	821		665	619		043	202	.70	798	375	12
49	254	858	.68	656	615	.06	075	244	.68	756	347	11
50	.32 292	9.50 896	.62	.94 646	9.97 610	.07	.34 108	9.53 285	.70	0.46 715	2.9319	10
51	309	933		637	606		140	327	.68	673	291	9
52	337	970		627	602	.08	173	368		632	263	8
53	364	9.51 007	.60	618	597	.07	205	409		591	235	7
54	392	043	.62	609	593		238	450	.70	550	208	6
55	419	080		599	589	.09	270	492	.66	508	180	5
56	447	117		590	584	.07	302	533		467	152	4
57	474	154		580	580		335	574		426	125	3
58	502	191	.60	571	576	.08	369	615		385	097	2
59	529	227	.62	561	571	.07	400	656		344	070	1
60	.32 557	9.51 264		.94 552	9.97 567		.34 433	9.53 697		0.46 303	2.9042	0

108°	Nat.	Log.	Dif.	Nat.	Log.	Dif.	Nat.	Log.	Dif.	Log.	Nat.	71°
		Cosines.			Sines.			Cotangents.			Tangents.	

19°	SINES.			COSINES.			TANGENTS.			COTANGENTS.		160°
	Nat.	Log.	Dif.	Nat.	Log.	Dif.	Nat.	Log.	Dif.	Log.	Nat.	
0′	.82 557	9.51 264	.62	.94 552	9.97 567	.07	.34 433	9.53 697	.68	0.46 303	2.9042	60′
1	584	301		542	563	.08	465	738		262	015	59
2	612	338	.60	533	558	.07	498	779		221	2.8087	58
3	630	374	.62	523	554		530	820		180	960	57
4	667	411	.60	514	550	.08	568	861		139	933	56
5	694	447	.62	504	545	.07	596	902		098	905	55
6	722	484	.60	495	541	.08	628	943		057	878	54
7	749	520	.62	485	536	.07	661	984		016	851	53
8	777	557	.60	476	532		693	9.54 025	.07	0.45 975	824	52
9	804	593		466	528	.08	726	065	.68	935	797	51
10	.82 832	9.51 629	.62	.94 457	9.97 523	.07	.34 758	9.54 106	.68	0.45 894	2.8770	50
11	859	666	.60	447	519		791	147	.67	853	748	49
12	887	702		438	515	.08	824	187	.08	813	716	48
13	914	738		428	510	.07	856	228		772	689	47
14	942	774	.62	418	506	.08	889	269	.07	731	662	46
15	969	811	.60	409	501	.07	922	309	.68	691	636	45
16	997	847		399	497	.08	954	350	.67	650	609	44
17	.83 024	883		390	492	.07	987	390	.68	610	582	43
18	051	919		380	488		.35 020	431	.67	569	556	42
19	079	955		370	484	.08	052	471	.68	529	529	41
20	.83 106	9.51 991	.60	.94 361	9.97 479	.07	.35 085	9.54 512	.67	0.45 488	2.8502	40
21	134	9.52 027		351	475	.08	119	552	.68	448	476	39
22	161	063		342	470	.07	150	593	.67	407	449	38
23	189	099		332	466	.08	183	633		367	423	37
24	216	135		322	461	.07	216	673	.08	327	397	36
25	244	171		313	457		248	714	.67	286	370	35
26	271	207	.58	303	453	.08	281	754		246	344	34
27	299	242	.60	293	448	.07	314	794	.68	206	318	33
28	326	278		284	444	.08	346	835	.67	165	291	32
29	353	314		274	439	.07	379	875		125	265	31
30	.33 881	9.52 350	.58	.94 264	9.97 435	.08	.35 412	9.54 915	.07	0.45 085	2.8239	30
31	408	385	.60	254	430		445	955		045	213	29
32	436	421	.59	245	426	.08	477	995		005	187	28
33	463	456	.60	235	421	.07	510	9.55 035		0.44 965	161	27
34	490	492	.58	225	417	.08	543	075		925	135	26
35	518	527	.60	215	412	.07	576	115		885	109	25
36	545	563	.58	206	408	.08	608	155		845	083	24
37	573	598	.60	196	403	.07	641	195		805	057	23
38	600	634	.58	186	399	.08	674	235		765	032	22
39	627	669	.60	176	394	.07	707	275		725	006	21
40	.33 655	9.52 705	.58	.94 167	9.97 390	.08	.35 740	9.55 315	.67	0.44 685	2.7960	20
41	682	740		157	385	.07	772	355	.05	645	955	19
42	710	775	.60	147	381	.08	805	395	.05	605	929	18
43	737	811	.58	137	376	.07	838	434	.07	566	903	17
44	764	846		127	372	.08	871	474		526	878	16
45	792	881		118	367	.07	904	514		486	852	15
46	819	916		108	363	.08	937	554	.05	446	827	14
47	846	951		098	358		969	593	.07	407	801	13
48	874	986		088	353	.07	.36 002	633		367	776	12
49	901	9.53 021		078	349	.08	035	673	.05	327	751	11
50	.33 929	9.53 056	.60	.94 068	9.97 344	.07	.36 068	9.55 712	.67	0.44 288	2.7725	10
51	956	092	.57	058	340	.08	101	752	.05	248	700	9
52	983	126	.58	049	335	.07	134	791	.67	209	675	8
53	.34 011	161		039	331	.08	167	831	.65	169	650	7
54	038	196		029	326	.07	199	870	.67	130	625	6
55	065	231		019	322	.08	232	910	.65	090	600	5
56	093	266		009	317		265	949	.67	051	575	4
57	120	301		.93 999	312	.07	298	989	.65	011	550	3
58	147	336	.57	989	308	.08	331	9.56 028		0.43 972	525	2
59	175	370	.58	979	303	.07	364	067	.67	933	500	1
60	.34 202	9.53 405		.93 969	9.97 299		.36 397	9.56 107		0.43 893	2.7475	0
109°	Nat.	Log.	Dif.	Nat.	Log.	Dif.	Nat.	Log.	Dif.	Log.	Nat.	70°
		COSINES.			SINES.			COTANGENTS.		TANGENTS.		

20°	SINES			COSINES			TANGENTS			COTANGENTS		159°
	Nat.	Log.	Dif.	Nat.	Log.	Dif.	Nat.	Log.	Dif.	Log.	Nat.	
0'	.34 202	9.53 405	.58	.93 969	9.97 299	.08	.36 397	9.56 107	.65	0.43 893	2.7475	60'
1	229	440		959	294		430	146		854	450	59
2	257	475	.57	949	289	.07	463	185		815	425	58
3	284	509	.58	939	285	.08	496	224	.67	776	400	57
4	311	544	.57	929	280	.07	529	264	.65	736	376	56
5	339	578	.58	919	276	.06	562	303		697	351	55
6	366	613	.57	909	271		595	342		658	326	54
7	393	647	.58	899	266	.07	628	381		619	302	53
8	421	682	.57	889	262	.06	661	420		580	277	52
9	448	716	.58	879	257		694	459		541	253	51
10	.34 475	9.53 751	.57	.93 869	9.97 252	.07	.36 727	9.56 498	.65	0.43 502	2.7228	50
11	503	785		859	248	.08	760	537		463	204	49
12	530	819	.58	849	243		793	576		424	179	48
13	557	854	.57	839	238	.07	826	615		385	155	47
14	584	888		829	234	.08	859	654		346	130	46
15	612	922	.58	819	229		892	693		307	106	45
16	639	957	.57	809	224	.07	925	732		268	082	44
17	666	991		799	220	.08	958	771		229	058	43
18	694	9.54 025		789	215		991	810		190	034	42
19	721	059		779	210	.07	.37 024	849	.63	151	009	41
20	.34 748	9.54 093	.57	.93 769	9.97 206	.08	.37 057	9.56 887	.65	0.43 113	2.6985	40
21	775	127		759	201		090	926		074	961	39
22	803	161		748	196	.07	123	965		035	937	38
23	830	195		738	192	.08	156	9.57 004	.63	0.42 996	913	37
24	857	229		728	187		189	042	.65	958	889	36
25	884	263		718	182	.07	223	081		919	865	35
26	912	297		708	178	.08	256	120	.63	880	841	34
27	939	331		698	173		289	158	.65	842	818	33
28	966	365		688	168		322	197	.63	803	794	32
29	993	399		677	163	.07	355	235	.65	765	770	31
30	.35 021	9.54 433	.55	.93 667	9.97 159	.08	.37 388	9.57 274	.63	0.42 726	2.6746	30
31	048	466	.57	657	154		422	312	.65	688	723	29
32	075	500		647	149	.07	455	351	.68	649	699	28
33	102	534	.55	637	145	.08	488	389	.65	611	676	27
34	130	567	.57	626	140		521	428	.63	572	652	26
35	157	601		616	135		554	466		534	628	25
36	184	635	.55	606	130	.07	588	504	.65	496	605	24
37	211	668	.57	596	126	.08	621	543	.63	457	581	23
38	239	702	.55	585	121		654	581		419	558	22
39	266	735	.57	575	116		687	619	.65	381	534	21
40	.35 293	9.54 769	.55	.93 565	9.97 111	.07	.37 720	9.57 658	.63	0.42 342	2.6511	20
41	320	802	.57	555	107	.08	754	696		304	488	19
42	347	836	.55	544	102		787	734		266	464	18
43	375	869	.57	534	097		820	772		228	441	17
44	402	903	.55	524	092		853	810	.65	190	418	16
45	429	936		514	087	.07	887	849	.63	151	395	15
46	456	969	.57	503	083	.08	920	887		113	371	14
47	484	9.55 003	.55	493	078		953	925		075	348	13
48	511	036		483	073		986	963		037	325	12
49	538	069		472	068		.38 020	9.58 001		0.41 999	302	11
50	.35 565	9.55 102	.57	.93 462	9.97 063	.07	.38 053	9.58 039	.63	0.41 961	2.6279	10
51	592	136	.55	452	059	.08	086	077		923	256	9
52	619	169		441	054		120	115		885	233	8
53	647	202		431	049		153	153		847	210	7
54	674	235		420	044		186	191		809	187	6
55	701	268		410	039	.07	220	229		771	165	5
56	728	301		400	035	.09	253	267	.62	733	142	4
57	755	334		389	030		286	304	.63	696	119	3
58	782	367		379	025		320	342		658	096	2
59	810	400		368	020		353	380		620	074	1
60	.35 837	9.55 433		.93 358	9.97 015		.38 386	9.58 418		0.41 582	2.6051	0

110°	Nat.	Log.	Dif.	Nat.	Log.	Dif.	Nat.	Log.	Dif.	Log.	Nat.	69°
		COSINES.			SINES.			COTANGENTS.			TANGENTS.	

21°	SINES			COSINES			TANGENTS			COTANGENTS		158°
	Nat.	Log.	Dif.	Nat.	Log.	Dif.	Nat.	Log.	Dif.	Log.	Nat.	
0'	.35 837	9.55 433	.55	.93 358	9.97 015	.08	.38 386	9.58 418	.62	0.41 582	2.6051	60'
1	864	466		348	010		420	455	.63	545	028	59
2	891	499		337	005	.07	453	493		507	006	58
3	918	532	.53	327	001	.08	487	531		469	2.5988	57
4	945	564	.55	316	9.96 996		520	569	.62	431	961	56
5	972	597		306	991		553	606	.63	394	938	55
6	.36 000	630		295	986		587	644	.62	356	916	54
7	027	663	.53	285	981		620	681	.63	319	898	53
8	054	695	.55	274	976		654	719		281	871	52
9	081	728		264	971		687	757	.62	243	848	51
10	.36 108	9.55 761	.53	.93 253	9.96 966	.07	.38 721	9.58 794	.63	0.41 206	2.5826	50
11	135	793	.55	243	962	.08	754	832	.62	168	804	49
12	162	826	.53	232	957		787	869	.63	131	782	48
13	190	858	.55	222	952		821	907	.62	093	759	47
14	217	891	.53	211	947		854	944		056	737	46
15	244	923	.55	201	942		888	981	.03	019	715	45
16	271	956	.53	190	937		921	9.59 019	.62	0.40 981	693	44
17	298	988	.55	180	932		955	056	.63	944	671	43
18	325	9.56 021	.53	169	927		988	094	.62	906	649	42
19	352	053		159	922		.39 022	131		869	627	41
20	.36 379	9.56 085	.55	.93 148	9.96 917	.08	.39 055	9.59 168	.62	0.40 832	2.5605	40
21	406	118	.53	137	912		089	205	.63	795	558	39
22	434	150		127	907	.07	122	243	.62	757	561	38
23	461	182	.55	116	903	.06	156	280		720	539	37
24	488	215	.53	106	898		190	317		683	517	36
25	515	247		095	893		223	354		646	495	35
26	542	279		084	888		257	391	.63	609	473	34
27	569	311		074	883		290	429	.62	571	452	33
28	596	343		063	878		324	466		534	430	32
29	623	375	.55	052	873		357	503		497	408	31
30	.36 650	9.56 408	.53	.93 042	9.96 868	.08	.39 391	9.59 540	.62	0.40 460	2.5386	30
31	677	440		031	863		425	577		423	365	29
32	704	472		020	858		458	614		386	343	28
33	731	504		010	853		492	651		349	322	27
34	758	536		.92 999	848		526	688		312	300	26
35	785	568	.52	988	843		559	725		275	279	25
36	812	599	.53	978	838		593	762		238	257	24
37	839	631		967	833		626	799	.60	201	236	23
38	867	663		956	828		660	835	.62	165	214	22
39	894	695		945	823		694	872		128	193	21
40	.36 921	9.56 727	.53	.92 935	9.96 818	.08	.39 727	9.59 909	.62	0.40 091	2.5172	20
41	948	759	.52	924	813		761	946		054	150	19
42	975	790	.53	913	808		795	983	.00	017	129	18
43	.37 002	822		903	803		829	9.60 019	.62	0.39 981	108	17
44	029	854		892	798		862	056		944	086	16
45	056	886	.52	881	793		896	093		907	065	15
46	083	917	.53	870	788		930	130	.60	870	044	14
47	110	949	.52	859	783		963	166	.62	834	023	13
48	137	980	.53	849	778	.10	997	203		797	002	12
49	164	9.57 012		838	772	.08	.40 031	240	.60	760	2.4961	11
50	.37 191	9.57 044	.52	.92 827	9.96 767	.08	.40 065	9.60 276	.62	0.39 724	2.4960	10
51	218	075	.53	816	762		098	313	.60	687	939	9
52	245	107	.52	805	757		132	349	.62	651	918	8
53	272	138		794	752		166	386	.60	614	897	7
54	299	169	.53	784	747		200	422	.62	578	876	6
55	326	201	.52	773	742		234	459	.60	541	855	5
56	353	232	.53	762	737		267	495	.62	505	834	4
57	380	264	.52	751	732		301	532	.60	468	818	3
58	407	295		740	727		335	568	.62	432	792	2
59	434	326	.53	729	722		369	605	.60	395	772	1
60	.37 461	9.57 358		.92 718	9.96 717		.40 408	9.60 641		0.39 359	2.4751	0

111°	Nat.	Log.	Dif.	Nat.	Log.	Dif.	Nat.	Log.	Dif.	Log.	Nat.	68°
		COSINES.			SINES.			COTANGENTS.		TANGENTS.		

22°	SINES.			COSINES.			TANGENTS.			COTANGENTS.		157°
	Nat.	Log.	Dif.	Nat.	Log.	Dif.	Nat.	Log.	Dif.	Log.	Nat.	
0'	.37 461	9.57 358	.52	.92 718	9.96 717	.10	.40 403	9.60 641	.60	0.39 359	2.4751	60'
1	488	389		707	711	.08	436	677	.62	323	730	59
2	515	420		697	706		470	714	.60	286	709	58
3	.542	451		686	701		504	750		250	689	57
4	569	482	.53	675	696		538	786	.62	214	668	56
5	595	514	.52	664	691		572	823	.60	177	648	55
6	622	545		658	686		606	859		141	627	54
7	649	576		642	681		640	895		105	606	53
8	676	607		631	676	.10	674	931		069	586	52
9	703	638		620	670	.08	707	967	.62	033	566	51
10	.37 730	9.57 669	.52	.92 609	9.96 665	.08	.40 741	9.61 004	.60	0.38 996	2.4545	50
11	757	700		598	660		775	040		960	525	49
12	784	731		587	655		809	076		924	504	48
13	811	762		576	650		843	112		888	484	47
14	838	793		565	645		877	148		852	464	46
15	865	824		554	640	.10	911	184		816	443	45
16	892	855	.50	543	634	.08	945	220		780	423	44
17	919	885	.52	532	629		979	256		744	408	43
18	946	916		521	624		.41 013	292		708	388	42
19	973	947		510	619		047	328		672	802	41
20	.37 999	9.57 978	.50	.92 499	9.96 614	.10	.41 081	9.61 364	.60	0.38 636	2.4342	40
21	.38 026	9.58 008	.52	488	608	.08	115	400		600	322	39
22	053	039		477	603		149	436		564	302	38
23	080	070		466	598		183	472		528	282	37
24	107	101	.50	455	593		217	508		492	262	36
25	134	131	.52	444	588	.10	251	544	.58	456	242	35
26	161	162	.50	432	582	.08	285	579	.60	421	222	34
27	188	192	.52	421	577		319	615		385	202	33
28	215	223	.50	410	572		353	651		349	182	32
29	241	253	.52	399	567		387	687	.58	313	162	31
30	.38 268	9.58 284	.50	.92 388	9.96 562	.10	.41 421	9.61 722	.60	0.38 278	2.4142	30
31	295	314	.52	377	556	.08	455	758		242	122	29
32	322	345	.50	366	551		490	794		206	102	28
33	349	375	.52	355	546		524	830	.58	170	083	27
34	376	406	.50	343	541	.10	558	865	.60	135	063	26
35	403	436	.52	332	535	.08	592	901	.58	099	043	25
36	430	467	.50	321	530		626	936	.60	064	023	24
37	456	497		310	525		660	972		028	004	23
38	483	527		299	520	.10	694	9.62 008	.58	0.37 992	2.3984	22
39	510	557	.52	287	514	.08	728	043	.60	957	964	21
40	.38 537	9.58 588	.50	.92 276	9.96 509	.08	.41 763	9.62 079	.58	0.37 921	2.3945	20
41	564	618		265	504	.10	797	114	.60	886	925	19
42	591	648		254	498	.08	831	150	.58	850	906	18
43	617	678	.52	243	493	.	865	185	.60	815	886	17
44	644	709	.50	231	488		899	221	.58	779	867	16
45	671	739		220	483	.10	933	256	.60	744	847	15
46	698	769		209	477	.08	968	292	.58	708	828	14
47	725	799		198	472		.42 002	327		673	808	13
48	752	829		186	467	.10	036	362	.60	638	789	12
49	778	859		175	461	.08	070	398	.58	602	770	11
50	.38 805	9.58 889	.50	.92 164	9.96 456	.08	.42 105	9.62 433	.58	0.37 567	2.3750	10
51	832	919		152	451	.10	139	468	.60	532	731	9
52	859	949		141	445	.08	173	504	.58	496	712	8
53	886	979		130	440		207	539		461	693	7
54	912	9.59 009		119	435	.10	242	574		426	673	6
55	939	039		107	429	.08	276	609	.60	391	654	5
56	966	069	.48	096	424		310	645	.58	355	635	4
57	993	098	.50	085	419	.10	345	680		320	616	3
58	.39 020	128		073	413	.08	379	715		285	597	2
59	046	158		062	408		413	750		250	578	1
60	.39 073	9.59 188		.92 050	9.96 403		.42 447	9.62 785		0.37 215	2.3559	0

112°	Nat.	Log.	Dif.	Nat.	Log.	Dif.	Nat.	Log.	Dif.	Log.	Nat.	67°
		COSINES.			SINES.			COTANGENTS.			TANGENTS.	

23°	SINES.			COSINES.			TANGENTS.			COTANGENTS.		156°
	Nat.	Log.	Dif.	Nat.	Log.	Dif.	Nat.	Log.	Dif.	Log.	Nat.	
0′	.39 078	9.59 188	.50	.92 050	9.96 403	.10	.42 447	9.62 785	.58	0.37 215	2.8550	60′
1	100	218	.48	080	397	.08	482	820		180	539	59
2	127	247	.50	028	392		516	855		145	520	58
3	153	277		016	387	.10	551	890	.60	110	501	57
4	180	307	.48	005	381	.08	585	926	.58	074	483	56
5	207	336	.50	.91 094	376	.10	619	961		039	464	55
6	284	366		982	370	.08	654	996		004	445	54
7	260	396	.48	971	365		688	9.63 031		0.36 969	426	53
8	287	425	.50	959	360	.10	722	066		934	407	52
9	314	455	.48	948	354	.08	757	101	.57	899	388	51
10	.39 841	9.59 484	.50	.91 936	9.96 349	.10	.42 791	9.63 135	.58	0.36 865	2.8369	50
11	867	514	.48	925	343	.08	826	170		830	351	49
12	894	543	.50	914	338		860	205		795	332	48
13	421	573	.48	902	333	.10	694	240		760	313	47
14	448	602	.50	891	327	.08	929	275		725	294	46
15	474	632	.48	879	322	.10	963	310		690	276	45
16	501	661		868	316	.08	998	345	.57	655	257	44
17	528	690	.50	856	311	.10	.43 032	379	.58	621	238	43
18	555	720	.48	845	305	.08	067	414		586	220	42
19	581	749		833	300	.10	101	449		551	201	41
20	.39 608	9.59 778	.50	.91 822	9.96 294	.08	.43 136	9.63 484	.58	0.36 516	2.8158	40
21	635	808	.48	810	289		170	519	.57	481	164	39
22	661	837		799	284	.10	205	553	.58	447	146	38
23	688	866		787	278	.08	289	588		412	127	37
24	715	895		775	273	.10	274	623	.57	377	109	36
25	741	924	.50	764	267	.08	308	657	.58	343	090	35
26	768	954	.48	752	262	.10	343	692	.57	308	072	34
27	795	983		741	256	.08	378	726	.58	274	053	33
28	822	9.60 012		729	251	.10	412	761		239	035	32
29	848	041		718	245	.08	447	796	.57	204	017	31
30	.39 875	9.60 070	.48	.91 706	9.96 240	.10	.43 451	9.63 830	.58	0.36 170	2.7998	30
31	902	099		694	234	.08	516	865	.57	135	980	29
32	928	128		683	229	.10	550	899	.58	101	962	28
33	955	157		671	223	.08	585	934	.57	066	944	27
34	982	186		660	218	.10	620	968	.58	032	925	26
35	.40 008	215		648	212	.08	654	9.64 003	.57	0.35 997	907	25
36	035	244		636	207	.10	689	037	.58	963	889	24
37	062	273		625	201	.08	724	072	.57	928	871	23
38	088	302		613	196	.10	758	106		894	853	22
39	115	331	.47	601	190	.08	793	140	.58	860	835	21
40	.40 141	9.60 359	.48	.91 590	9.96 185	.10	.43 828	9.64 175	.57	0.35 825	2.7817	20
41	168	388		578	179	.08	862	209		791	790	19
42	195	417		566	174	.10	897	243	.58	757	781	18
43	221	446	.47	555	168		932	278	.57	722	763	17
44	248	474	.48	543	162	.08	966	312		688	745	16
45	275	503		531	157	.10	.44 001	346	.58	654	727	15
46	301	532		519	151	.08	086	381	.57	619	709	14
47	328	561	.47	508	146	.10	071	415		585	691	13
48	355	589	.48	496	140	.08	105	449		551	673	12
49	381	618	.47	484	135	.10	140	483		517	655	11
50	.40 408	9.60 646	.48	.91 472	9.96 129	.10	.44 175	9.64 517	.58	0.35 483	2.7637	10
51	434	675		461	123	.08	210	552	.57	448	620	9
52	461	704	.47	449	118	.10	244	586		414	602	8
53	488	732	.48	437	112	.08	279	620		380	584	7
54	514	761	.47	425	107	.10	314	654		346	566	6
55	541	789	.48	414	101		349	688		312	549	5
56	567	818	.47	402	095	.08	384	722		278	531	4
57	594	846	.48	390	090	.10	418	756		244	513	3
58	621	875	.47	378	084	.08	453	790		210	496	2
59	647	903	.47	366	079	.10	488	824		176	478	1
60	.40 674	9.60 931		.91 355	9.96 073		.44 523	9.64 858		0.35 142	2.2460	0

113°	Nat.	Log.	Dif.	Nat.	Log.	Dif.	Nat.	Log.	Dif.	Log.	Nat.	66°
		COSINES.			SINES.			COTANGENTS.			TANGENTS.	

24°	SINES			COSINES			TANGENTS			COTANGENTS		155°
	Nat.	Log.	Dif.	Nat.	Log.	Dif.	Nat.	Log.	Dif.	Log.	Nat.	
0'	.40 674	9.60 931	.48	.91 855	9.96 073	.10	.44 523	9.64 858	.57	0.35 142	2.2460	60'
1	700	960	.47	843	067	.08	558	892		108	443	59
2	727	988		831	062	.10	593	926		074	425	58
3	753	9.61 016	.48	819	056		627	960		040	408	57
4	780	045	.47	307	050	.08	662	994		006	390	56
5	806	073		295	045	.10	697	9.65 028		0.34 972	378	55
6	833	101		283	039	.08	732	062		938	855	54
7	860	129	.48	272	034	.10	767	096		904	838	53
8	886	158	.47	260	028		802	130		870	820	52
9	913	186		248	022	.08	837	164	.55	836	803	51
10	.40 939	9.61 214	.47	.91 236	9.96 017	.10	.44 872	9.65 197	.57	0.34 803	2.2286	50
11	966	242		224	011		907	231		769	268	49
12	992	270		212	005	.08	942	265		735	251	48
13	.41 019	298		200	000	.10	977	299		701	234	47
14	045	326		188	9.95 994		.45 012	333	.55	667	216	46
15	072	354		176	988		047	366	.57	634	199	45
16	098	382	.48	164	982	.08	092	400		600	182	44
17	125	411	.45	152	977	.10	117	434	.55	566	165	43
18	151	438	.47	140	971		152	467	.57	533	148	42
19	178	466		128	965	.08	187	501		499	130	41
20	.41 204	9.61 494	.47	.91 116	9.95 960	.10	.45 222	9.65 535	.55	0.34 465	2.2113	40
21	231	522		104	954		257	568	.57	432	096	39
22	257	550		092	948		292	602		398	079	38
23	284	573		080	942	.08	327	636	.55	364	062	37
24	310	606		068	937	.10	362	669	.57	331	045	36
25	337	634		056	931		397	703	.55	297	028	35
26	363	662	.45	044	925	.08	432	736	.57	264	011	34
27	390	689	.47	032	920	.10	467	770	.55	230	2.1994	33
28	416	717		020	914		502	803	.57	197	977	32
29	443	745		008	908		538	837	.55	163	960	31
30	.41 469	9.61 773	.45	.90 996	9.95 902	.09	.45 573	9.65 870	.57	0.34 130	2.1943	30
31	496	800	.47	984	897	.10	608	904	.55	096	926	29
32	522	828		972	891		643	937	.57	063	909	28
33	549	856	.45	960	885		678	971	.55	029	892	27
34	575	883	.47	948	879		713	9.66 004	.57	0.33 996	876	26
35	602	911		936	873	.08	748	038	.55	962	859	25
36	628	939	.45	924	868	.10	784	071		929	842	24
37	655	966	.47	911	862		819	104	.57	896	825	23
38	681	994	.45	899	856		854	138	.55	862	808	22
39	707	9.62 021	.47	887	850		889	171		829	792	21
40	.41 734	9.62 049		.90 875	9.95 844	.08	.45 924	9.66 204	.57	0.33 796	2.1775	20
41	760	076	.47	863	839	.10	960	238	.55	762	758	19
42	787	104	.45	851	833		995	271		729	742	18
43	813	131	.45	839	827		.46 030	304		696	725	17
44	840	159	.45	826	821		065	337	.57	663	708	16
45	866	186	.47	814	815	.08	101	371	.55	629	692	15
46	892	214	.45	802	810	.10	136	404		596	675	14
47	919	241		790	804		171	437		563	659	13
48	945	268	.47	778	798		206	470		530	642	12
49	972	296	.45	766	792		242	503	.57	497	625	11
50	.41 998	9.62 323	.45	.90 758	9.95 786	.10	.46 277	9.66 537	.55	0.33 463	2.1600	10
51	.42 024	350		741	780	.08	312	570		430	592	9
52	051	377	.47	729	775	.10	348	603		397	576	8
53	077	405	.45	717	769		383	636		364	560	7
54	104	432		704	763		418	669		331	543	6
55	130	459		692	757		454	702		298	527	5
56	156	486		680	751		489	735		265	510	4
57	183	513	.47	668	745		525	768		232	494	3
58	209	541	.45	655	739		560	801		199	478	2
59	235	568		643	733	.09	595	834		166	461	1
60	.42 262	9.62 595		.90 631	9.95 728		.46 631	9.66 867		0.33 133	2.1445	0

114°	Nat.	Log.	Dif.	Nat.	Log.	Dif.	Nat.	Log.	Dif.	Log.	Nat.	65°
		COSINES.			SINES.			COTANGENTS.			TANGENTS.	

25°	Sines.			Cosines.			Tangents.			Cotangents.		154°
	Nat.	Log.	Dif.	Nat.	Log.	Dif.	Nat.	Log.	Dif.	Log.	Nat.	
0'	.42 262	9.62 595	.45	.90 631	9.95 728	.10	.46 631	9.66 867	.55	0.33 133	2.1445	60'
1	288	622		618	722		666	900		100	429	59
2	315	649		606	716		702	933		067	418	58
3	341	676		594	710		737	966		034	396	57
4	367	703		582	704		772	999		001	380	56
5	394	730		569	698		808	9.67 032		0.32 968	864	55
6	420	757		557	692		843	065		935	348	54
7	446	784		545	686		879	098		902	332	53
8	473	811		532	680		914	131	.56	869	315	52
9	499	838		520	674		950	163	.55	837	299	51
10	.42 525	9.62 865	.45	.90 507	9.95 668	.08	.46 985	9.67 196	.55	0.32 804	2.1283	50
11	552	892	.43	495	663	.10	.47 021	229		771	267	49
12	578	918	.45	483	657		056	262		738	251	48
13	604	945		470	651		092	295	.58	705	285	47
14	631	972		458	645		128	327	.55	673	219	46
15	657	999		446	639		163	360		640	208	45
16	683	9.63 026	.48	433	633		199	393		607	187	44
17	709	052	.45	421	627		234	426	.53	574	171	43
18	736	079		408	621		270	458	.55	542	155	42
19	762	106		396	615		305	491		509	189	41
20	.42 788	9.63 133	.48	.90 383	9.95 609	.10	.47 341	9.67 524	.53	0.32 476	2.1123	40
21	815	159	.45	371	603		377	556	.55	444	107	39
22	841	186		358	597		412	589		411	092	38
23	867	213	.48	346	591		448	622	.58	378	076	37
24	894	239	.45	334	585		483	654	.55	346	060	36
25	920	266	.48	321	579		519	687	.53	313	044	35
26	946	292	.45	309	573		555	719	.55	281	028	34
27	972	319	.48	296	567		590	752		248	013	33
28	999	345	.45	284	561		626	785	.53	215	2.0997	32
29	.43 025	372	.48	271	555		662	817	.55	183	981	31
30	.43 051	9.63 398	.45	.90 259	9.95 549	.10	.47 698	9.67 850	.53	0.32 150	2.0965	30
31	077	425	.48	246	543		733	882	.55	118	950	29
32	104	451	.45	233	537		769	915	.58	085	934	28
33	130	478	.48	221	531		805	947	.55	053	916	27
34	156	504	.45	209	525		840	980	.58	020	903	26
35	182	531	.48	196	519		876	9.68 012		0.31 988	887	25
36	209	557		183	513		912	044	.55	956	872	24
37	235	583	.45	171	507	.12	948	077	.53	923	856	23
38	261	610	.48	158	500	.10	984	109	.55	891	840	22
39	287	636		146	494		.48 019	142	.59	858	825	21
40	.43 313	9.63 662	.45	.90 133	9.95 488	.10	.48 055	9.68 174	.58	0.31 826	2.0609	20
41	340	689	.48	120	482		091	206	.55	794	794	19
42	366	715		108	476		127	239	.58	761	778	18
43	392	741		095	470		163	271		729	763	17
44	418	767	.45	082	464		198	303	.55	697	748	16
45	445	794	.48	070	458		234	336	.58	664	732	15
46	471	820		057	452		270	368		632	717	14
47	497	846		045	446		306	400		600	701	13
48	523	872		032	440		342	432	.55	568	686	12
49	549	898		020	434	.12	378	465		535	671	11
50	.43 575	9.63 924	.48	.90 007	9.95 427	.10	.48 414	9.68 497	.58	0.31 503	2.0655	10
51	602	950		.89 994	421		450	529		471	640	9
52	628	976		981	415		486	561		439	625	8
53	654	9.64 002		968	409		521	593	.55	407	609	7
54	680	028		956	403		557	626	.58	374	594	6
55	706	054		943	397		593	658		342	579	5
56	733	080		930	391	.12	629	690		310	564	4
57	759	106		918	384	.10	665	722		278	549	3
58	785	132		905	378		701	754		246	533	2
59	811	158		892	372		737	786		214	519	1
60	.43 837	9.64 184		.89 879	9.95 366		.48 773	9.68 818		0.31 182	2.0503	0
115°	Nat.	Log.	Dif.	Nat.	Log.	Dif.	Nat.	Log.	Dif.	Log.	Nat.	64°
		Cosines.			Sines.			Cotangents.		Tangents.		

26°	Sines			Cosines			Tangents			Cotangents		153°
	Nat.	Log.	Dif.	Nat.	Log.	Dif.	Nat.	Log.	Dif.	Log.	Nat.	
0'	.43 837	9.64 184	.43	.89 879	9.95 366	.10	.48 773	9.68 818	.53	0.31 182	2.0503	60'
1	863	210		867	360		809	850		150	488	59
2	889	236		854	354		845	882		118	473	58
3	916	262		841	348	.12	881	914		086	458	57
4	942	288	.42	828	341	.10	917	946		054	443	56
5	968	313	.43	816	335		953	978		022	428	55
6	994	339		803	329		980	9.69 010		0.30 990	413	54
7	.44 020	365		790	323		.49 026	042		958	898	53
8	046	391		777	317	.12	062	074		926	388	52
9	072	417	.42	764	310	.10	098	106		894	388	51
10	.44 098	9.64 442	.43	.89 752	9.95 304	.10	.49 134	9.69 138	.53	0.30 862	2.0353	50
11	124	468		739	298		170	170		830	838	49
12	151	494	.42	726	292		206	202		798	323	48
13	177	519	.43	713	286	.12	242	234		766	308	47
14	203	545		700	279	.10	278	266		734	293	46
15	229	571	.42	687	273		315	298	.52	702	278	45
16	255	596	.43	674	267		351	329	.53	671	268	44
17	281	622	.42	662	261	.12	387	361		639	248	43
18	307	647	.43	649	254	.10	423	393		607	238	42
19	333	673	.42	636	248		459	425		575	210	41
20	.44 359	9.64 698	.43	.89 623	9.95 242	.10	.49 495	9.69 457	.52	0.30 543	2.0204	40
21	385	724	.42	610	236	.12	532	488	.53	512	189	39
22	411	749	.43	597	229	.10	568	520		480	174	38
23	437	775	.42	584	223		604	552		448	160	37
24	464	800	.43	571	217		640	584	.52	416	145	36
25	490	826	.42	558	211	.12	677	615	.53	385	130	35
26	516	851	.43	545	204	.10	713	647		353	115	34
27	542	877	.42	532	198		749	679	.52	321	101	33
28	568	902	.43	519	192	.12	786	710	.56	290	086	32
29	594	927	.43	506	185	.10	822	742		258	072	31
30	.44 620	9.64 953	.42	.89 493	9.95 179	.10	.49 858	9.69 774	.52	0.30 226	2.0057	30
31	646	978		480	173		894	805	.53	195	042	29
32	672	9.65 003	.43	467	167	.12	931	837	.52	163	028	28
33	698	029	.42	454	160	.10	967	868	.53	132	013	27
34	724	054		441	154		.50 004	900		100	1.9999	26
35	750	079		428	148	.12	040	932	.52	068	984	25
36	776	104	.43	415	141	.10	076	963	.53	037	970	24
37	802	130	.42	402	135		113	995	.52	005	955	23
38	828	155		389	129	.12	149	9.70 026	.53	0.29 974	941	22
39	854	180		376	122	.10	185	058	.52	942	926	21
40	.44 880	9.65 205	.42	.89 363	9.95 116	.10	.50 222	9.70 089	.53	0.29 911	1.9912	20
41	906	230		350	110	.12	258	121	.52	879	897	19
42	932	255	.43	337	103	.10	295	152	.53	848	883	18
43	958	281	.42	324	097	.12	331	184	.52	816	868	17
44	984	306		311	090	.10	368	215	.53	785	854	16
45	.45 010	331		298	084		404	247	.52	753	840	15
46	036	356		285	078	.12	441	278		722	825	14
47	062	381		272	071	.10	477	309	.53	691	811	13
48	088	406		259	065		514	341	.52	659	797	12
49	114	431		245	059	.12	550	372	.53	628	782	11
50	.45 140	9.65 456	.42	.89 232	9.95 052	.10	.50 587	9.70 404	.52	0.29 596	1.9768	10
51	166	481		219	046	.12	623	435		565	754	9
52	192	506		206	039	.10	660	466	.53	534	740	8
53	218	531		193	033		696	498	.52	502	725	7
54	244	556	.40	180	027	.12	733	529		471	711	6
55	269	580	.42	167	020	.10	769	560	.53	440	697	5
56	295	605		153	014	.12	806	592	.52	408	683	4
57	321	630		140	007	.10	843	623		377	669	3
58	347	655		127	001		879	654		346	654	2
59	373	680		114	9.94 995	.12	916	685	.53	315	640	1
60	.45 399	9.65 705		.89 101	9.94 988		.50 953	9.70 717		0.29 283	1.9626	0
116°	Nat.	Log.	Dif.	Nat.	Log.	Dif.	Nat.	Log.	Dif.	Log.	Nat.	63°
		Cosines.			Sines.			Cotangents.			Tangents.	

27°	SINES.			COSINES.			TANGENTS.			COTANGENTS.		152°
	Nat.	Log.	Dif.	Nat.	Log.	Dif.	Nat.	Log.	Dif.	Log.	Nat.	
0'	.45 399	9.65 705	.40	.89 101	9.94 988	.10	.50 953	9.70 717	.52	0.29 283	1.9626	60'
1	425	729	.42	087	982	.12	989	748		252	612	59
2	451	754		074	975	.10	.51 026	779		.221	598	58
3	477	779		061	969	.12	068	810		190	584	57
4	503	804	.40	048	962	.10	099	841	.53	159	570	56
5	520	828	.42	035	956	..12	136	873	.52	127	556	55
6	554	853		021	949	.10	173	904		096	542	54
7	580	878	.40	008	943	.12	209	935		065	528	53
8	606,	902	.42	.88 995	936	.10	246	966		034	514	52
9	632	927		981	930	.12	283	997		003	500	51
10	.45 658	9.65 952	.40	.88 968	9.94 923	.10	.51 319	9.71 028	.52	0.28 972	1.9486	50
11	684	976	.42	955	917		356	059		941	472	49
12	710	9.66 001	.40	942	911	.12	393	090		910	458	48
13	736	025	.42	928	904	.10	430	121	.53	879	444	47
14	762	050		915	898	.12	467	153	.52	847	430	46
15	787	075	.40	902	891	.10	503	184		816	416	45
16	813	099	.42	888	885	.12	540	215		785	402	44
17	839	124	.40	875	878		577	246		754	388	43
18	865	148	.42	862	871	.10	614	277		723	375	42
19	891	173	.40	848	865	.12	651	308		692	361	41
20	.45 917	9.66 197	.40	.88 835	9.94 858	.10	.51 688	9.71 339	.52	0.28 661	1.9347	40
21	942	221	.42	822	852	.12	724	370		630	333	39
22	968	246	.40	808	845	.10	761	401	.50	599	319	38
23	994	270	.42	795	839	.12	798	431	.52	569	306	37
24	.46 020	295	.40	792	832	.10	835	462		538	292	36
25	046	319		768	826	.12	872	493		507	278	35
26	072	343	.42	755	819	.10	909	524		476	265	34
27	097	368	.40	741	813	.12	946	555		445	251	33
28	123	392		728	806		983	586		414	287	32
29	149	416	.42	715	799	.10	.52 020	617		383	228	31
30	.46 175	9.66 441	.40	.88 701	9.94 793	.12	.52 057	9.71 648	.52	0.28 352	1.9210	30
31	201	465		688	786	.10	094	679	.50	321	196	29
32	226	489		674	780	.12	131	709	.52	291	183	28
33	252	513		661	773	.10	168	740		260	169	27
34	278	537	.42	647	767	.12	205	771		229	155	26
35	304	562	.40	634	760		242	802		198	142	25
36	330	586		620	753	.10	279	833	.50	167	128	24
37	355	610		607	747	.12	316	863	.52	137	115	23
38	381	634		593	740	.10	353	894		106	101	22
39	407	658		580	734	.12	390	925	.50	075	088	21
40	.46 433	9.66 682	.40	.88 566	9.94 727	.12	.52 427	9.71 955	.52	0.28 045	1.9074	20
41	458	706	.42	553	720	.10	464	986		014	061	19
42	484	731	.40	539	714	.12	501	9.72 017		0.27 983	047	18
43	510	755		526	707		538	048	.50	952	034	17
44	536	779		512	700	.10	575	078	.52	922	020	16
45	561	803		499	694	.12	613	109		891	007	15
46	587	827		485	687		650	140	.50	860	1.8993	14
47	613	851		472	680	.10	687	170	.52	830	980	13
48	639	875		458	674	.12	724	201	.50	799	967	12
49	664	899	.38	445	667		761	231	.52	769	953	11
50	.46 690	9.66 922	.40	.88 431	9.94 660	.10	.52 798	9.72 262	.52	0.27 738	1.8940	10
51	716	946		417	654	.12	836	293	.50	707	927	9
52	742	970		404	647		873	323	.52	677	913	8
53	767	994		390	640	.10	910	354	.50	646	900	7
54	793	9.67 018		377	634	.12	947	384	.52	616	887	6
55	819	042		363	627		985	415	.50	585	873	5
56	844	066		349	620	.10	.53 022	445	.50	555	860	4
57	870	090	.38	336	614	.12	059	476	.50	524	847	3
58	896	113	.40	322	607		096	506	.52	494	834	2
59	921	137		308	600		134	537	.50	463	820	1
60	.46 947	9.67 161		.88 295	9.94 593		.53 171	9.72 567		0.27 433	1.8807	0
117°	Nat.	Log.	Dif.	Nat.	Log.	Dif.	Nat.	Log.	Dif.	Log.	Nat.	62°
		COSINES.			SINES.			COTANGENTS.			TANGENTS.	

28°	SINES.			COSINES.			TANGENTS.			COTANGENTS.		151°
	Nat.	Log.	Dif.	Nat.	Log.	Dif.	Nat.	Log.	Dif.	Log.	Nat.	
0′	.46 947	9.67 161	.40	.88 295	9.94 593	.10	.53 171	9.72 567	.52	0.27 433	1.8607	60′
1	973	185	.38	281	587	.12	208	598	.50	402	794	59
2	999	208	.40	267	580		246	628	.52	372	761	58
3	.47 024	232		254	573	.10	288	659	.50	341	768	57
4	050	256		240	567	.12	320	689	.52	311	735	56
5	076	280	.38	226	560		358	720	.50	280	741	55
6	101	303	.40	213	553		395	750		250	728	54
7	127	327	.88	199	546	.10	432	780	.52	220	715	53
8	153	350	.40	185	540	.12	470	811	.50	189	702	52
9	178	374		172	533		507	841	.52	159	689	51
10	.47 204	9.67 398	.38	.88 158	9.94 526	.12	.53 545	9.72 872	.50	0.27 128	1.8676	50
11	229	421	.40	144	519	.10	582	902		098	663	49
12	255	445	.38	130	513	.12	620	932	.52	068	650	48
13	281	468	.40	117	506		657	963	.50	037	637	47
14	306	492	.38	103	499		694	993		007	624	46
15	332	515	.40	089	492		732	9.73 023	.52	0.26 977	611	45
16	358	539	.38	075	485	.10	769	054	.50	946	598	44
17	383	562	.40	062	479	.12	807	084		916	585	43
18	409	586	.38	048	472		844	114		886	572	42
19	434	609	.40	034	465		882	144	.52	856	559	41
20	.47 460	9.67 633	.38	.88 020	9.94 458	.12	.53 920	9.73 175	.50	0.26 825	1.8546	40
21	486	656	.40	006	451	.10	957	205		795	533	39
22	511	680	.38	.87 993	445	.12	995	235		765	520	38
23	537	703		979	438		.54 032	265		735	507	37
24	562	726	.40	965	431		070	295	.52	705	495	36
25	588	750	.38	951	424		107	326	.50	674	482	35
26	614	773		937	417		145	356		644	469	34
27	639	796	.40	923	410	.10	183	386		614	456	33
28	665	820	.38	909	404	.12	220	416		584	443	32
29	690	843		896	397		258	446		554	480	31
30	.47 716	9.67 866	.40	.87 882	9.94 390	.12	.54 296	9.73 476	.52	0.26 524	1.8418	30
31	741	890	.88	868	383		333	507	.50	493	405	29
32	767	913		854	376		371	537		463	392	28
33	793	936		840	369		409	567		433	379	27
34	818	959		826	362		446	597		403	367	26
35	844	982	.40	812	355	.10	484	627		373	354	25
36	869	9.68 006	.88	798	349	.12	522	657		343	341	24
37	895	029		784	342		560	687		313	329	23
38	920	052		770	335		597	717		283	316	22
39	946	075		756	328		635	747		253	308	21
40	.47 971	9.68 098	.38	.57 743	9.94 321	.12	.54 673	9.73 777	.50	0.26 223	1.8291	20
41	997	121		729	314		711	807		193	278	19
42	.48 022	144		715	307		746	837		163	265	18
43	048	167		701	300		786	867		133	253	17
44	073	190		687	293		824	897		103	240	16
45	099	213	.40	673	286		862	927		073	228	15
46	124	237	.88	659	279	.10	900	957		043	215	14
47	150	260		645	273	.12	938	987		013	202	13
48	175	283	.37	631	266		975	9.74 017		0.25 983	190	12
49	201	305	.38	617	259		.55 013	047		953	177	11
50	.48 226	9.68 328	.88	.87 603	9.94 252	.12	.55 051	9.74 077	.50	0.25 923	1.8165	10
51	252	351		589	245		089	107		893	152	9
52	277	374		575	238		127	137	.48	863	140	8
53	303	397		561	231		165	166	.50	834	127	7
54	328	420		546	224		203	196		804	115	6
55	354	443		532	217		241	226		774	103	5
56	379	466		518	210		279	256		744	090	4
57	405	489		504	203		317	286		714	078	3
58	430	512	.37	490	196		355	316	.48	684	065	2
59	456	534	.38	476	189		393	345	.50	655	053	1
60	.48 481	9.68 557		.87 462	9.94 182		.55 431	9.74 375		0.25 625	1.8040	0
118°	Nat.	Log.	Dif.	Nat.	Log.	Dif.	Nat.	Log.	Dif.	Log.	Nat.	61°
		COSINES.			SINES.			COTANGENTS.			TANGENTS.	

29°	SINES			COSINES			TANGENTS			COTANGENTS		150°
′	Nat.	Log.	Dif.	Nat.	Log.	Dif.	Nat.	Log.	Dif.	Log.	Nat.	
0′	.48 481	9.68 557	.88	.87 462	9.94 182	.12	.55 481	9.74 375	.50	0.25 625	1.8040	60′
1	506	580		448	175		469	405		595	028	59
2	532	603	.87	434	168		507	435		565	016	58
3	557	625	.88	420	161		545	465	.48	535	008	57
4	583	648		406	154		583	494	.50	506	1.7991	56
5	608	671		391	147		621	524		476	979	55
6	634	694	.87	377	140		659	554	.48	446	966	54
7	659	716	.88	363	133		697	583	.50	417	954	53
8	684	739		349	126		736	613		387	942	52
9	710	762	.87	335	119		774	643		357	930	51
10	.48 735	9.68 784	.86	.87 321	9.94 112	.12	.55 812	9.74 673	.48	0.25 327	1.7917	50
11	761	807	.87	306	105		850	702	.50	298	905	49
12	786	829	.88	292	098	.13	888	732		268	893	48
13	811	852		278	090	.12	926	762	.48	238	881	47
14	837	875	.87	264	083		964	791	.50	209	868	46
15	862	897	.88	250	076		.56 008	821		179	856	45
16	888	920	.87	235	069		041	851	.48	149	844	44
17	913	942	.88	221	062		079	880	.50	120	832	43
18	938	965	.87	207	055		117	910	.48	090	820	42
19	964	987	.88	193	048		156	939	.50	061	808	41
20	.48 989	9.69 010	.87	.87 178	9.94 041	.12	.56 194	9.74 969	.48	0.25 031	1.7796	40
21	.49 014	032	.88	164	034		232	998	.50	002	783	39
22	040	055	.87	150	027		270	9.75 028		0.24 972	771	38
23	065	077	.88	136	020	.13	309	058	.48	942	759	37
24	090	100	.87	121	012	.12	347	087	.50	913	747	36
25	116	122		107	005		385	117	.48	883	735	35
26	141	144	.88	093	9.93 998		424	146	.50	854	723	34
27	166	167	.87	079	991		462	176	.48	824	711	33
28	192	189	.86	064	984		501	205	.50	795	699	32
29	217	212	.87	050	977		539	235	.48	765	687	31
30	.49 242	9.69 234	.87	.87 036	9.93 970	.12	.56 577	9.75 264	.50	0.24 736	1.7675	30
31	268	256	.88	021	963	.13	616	294	.48	706	663	29
32	293	279	.87	007	955	.12	654	323	.50	677	651	28
33	318	301		.86 993	948		693	353	.48	647	639	27
34	344	323		978	941		731	382		618	627	26
35	369	345	.88	964	934		769	411	.50	589	615	25
36	394	368	.87	949	927		808	441	.48	559	603	24
37	419	390		935	920	.13	846	470	.50	530	591	23
38	445	412		921	912	.12	885	500	.48	500	579	22
39	470	434		906	905		923	529		471	567	21
40	.49 495	9.69 456	.88	.86 892	9.93 898	.12	.56 962	9.75 558	.50	0.24 442	1.7556	20
41	521	479	.87	878	891		.57 000	588	.48	412	544	19
42	546	501		863	884	.13	039	617	.50	383	532	18
43	571	523		849	876	.12	078	647	.48	353	520	17
44	596	545		834	869		116	676		324	508	16
45	622	567		820	862		155	705	.50	295	496	15
46	647	589		805	855	.13	198	735	.48	265	485	14
47	672	611		791	847	.12	232	764		236	473	13
48	697	633		777	840		271	793		207	461	12
49	722	655		762	833		309	822	.50	178	449	11
50	.49 748	9.69 677	.87	.86 748	9.93 826	.12	.57 348	9.75 852	.48	0.24 148	1.7437	10
51	773	699		733	819	.13	386	881		119	426	9
52	798	721		719	811	.12	425	910		090	414	8
53	824	743		704	804		464	939	.50	061	402	7
54	849	765		690	797	.13	502	969	.48	031	391	6
55	874	787		675	789	.12	541	998		002	379	5
56	899	809		661	782		580	9.76 027		0.23 973	367	4
57	924	831		646	775		619	056	.50	944	355	3
58	950	853		632	768	.13	657	086	.48	914	344	2
59	975	875		617	760	.12	696	115		885	332	1
60	.50 000	9.69 897		.86 603	9.93 753		.57 735	9.76 144		0.23 856	1.7321	0

119°	Nat.	Log.	Dif.	Nat.	Log.	Dif.	Nat.	Log.	Dif.	Log.	Nat.	60°
		COSINES.			SINES.			COTANGENTS.			TANGENTS.	

30°	Sines. Nat.	Log.	Dif.	Cosines. Nat.	Log.	Dif.	Tangents. Nat.	Log.	Dif.	Cotangents. Log.	Nat.	149°
0'	.50 000	9.69 897	.37	.86 606	9.93 753	.12	.57 735	9.76 144	.48	0.23 856	1.7321	60'
1	025	919		588	746	.13	774	173		827	309	59
2	050	941		578	738	.12	818	202		798	297	58
3	076	963	.85	559	731		851	231	.50	769	286	57
4	101	984	.87	544	724		890	261	.48	739	274	56
5	126	9.70 006		530	717	.13	929	290		710	262	55
6	151	028		515	709	.12	968	319		681	251	54
7	176	050		501	702		.58 007	348		652	239	53
8	201	072	.85	486	695	.13	046	377		623	228	52
9	227	093	.87	471	687	.12	085	406		594	216	51
10	.50 252	9.70 115	.37	.86 457	9.93 680	.12	.58 124	9.76 435	.48	0.23 565	1.7205	50
11	277	137		442	673	.13	162	464		536	193	49
12	302	159	.85	427	665	.12	201	493		507	182	48
13	327	180	.37	413	658	.13	240	522		478	170	47
14	352	202		398	650	.12	279	551		449	159	46
15	377	224	.85	384	643		318	580		420	147	45
16	403	245	.37	369	636	.13	357	609	.50	391	136	44
17	428	267	.85	354	628	.12	396	639	.48	361	124	43
18	453	288	.37	340	621		485	668		332	113	42
19	478	310		325	614	.13	474	697	.47	303	102	41
20	.50 503	9.70 332	.85	.86 310	9.93 606	.12	.58 518	9.76 725	.48	0.23 275	1.7090	40
21	528	353	.87	295	599	.13	532	754		246	079	39
22	553	375	.85	281	591	.12	591	783		217	067	38
23	578	396	.87	266	584		631	812		188	056	37
24	608	418	.85	251	577	.13	670	841		159	045	36
25	628	439	.87	237	569	.12	709	870		130	033	35
26	654	461	.85	222	562	.13	748	899		101	022	34
27	679	482	.87	207	554	.12	787	928		072	011	33
28	704	504	.85	192	547	.13	826	957		043	1.6999	32
29	729	525	.87	178	539	.12	865	986		014	988	31
30	.50 754	9.70 547	.87	.86 163	9.93 532	.12	.58 905	9.77 015	.48	0.22 985	1.6977	30
31	779	568	.87	148	525	.13	944	044		956	965	29
32	804	590	.85	133	517	.12	983	073	.47	927	954	28
33	829	611	.87	119	510	.13	.59 022	101	.48	899	943	27
34	854	633	.85	104	502	.12	061	130		870	932	26
35	879	654		089	495	.13	101	159		841	920	25
36	904	675	.87	074	487	.12	140	188		812	909	24
37	929	697	.85	059	480	.13	179	217		783	898	23
38	954	718		045	472	.12	219	246	.47	754	887	22
39	979	739	.87	030	465	.13	258	274	.48	726	875	21
40	.51 004	9.70 761	.85	.86 015	9.93 457	.12	.59 297	9.77 303	.48	0.22 697	1.6864	20
41	029	782		000	450	.13	336	332		668	853	19
42	054	803		.85 985	442	.12	376	361		639	842	18
43	079	824	.87	970	435	.13	415	390	.47	610	831	17
44	104	846	.85	956	427	.12	454	418	.48	582	820	16
45	129	867		941	420	.13	494	447		553	808	15
46	154	888		926	412	.12	533	476		524	797	14
47	179	909	.87	911	405	.13	578	505	.47	495	786	13
48	204	931	.85	896	397	.12	612	533	.48	467	775	12
49	229	952		881	390	.13	651	562		438	764	11
50	.51 254	9.70 973	.85	.85 866	9.93 382	.12	.59 691	9.77 591	.47	0.22 409	1.6753	10
51	279	994		851	375	.13	730	619	.48	381	742	9
52	304	9.71 015		886	367	.12	770	648		352	731	8
53	329	036	.87	821	360	.13	809	677		323	720	7
54	354	058	.85	806	352		849	706	.47	294	709	6
55	379	079		792	344	.12	888	734	.48	266	698	5
56	404	100		777	337	.13	928	763	.47	237	687	4
57	429	121		762	329	.12	967	791	.48	209	676	3
58	454	142		747	322	.13	.60 007	820		180	665	2
59	479	163		732	314	.12	046	849	.47	151	654	1
60	.51 504	9.71 184		.85 717	9.93 307		.60 086	9.77 877		0.22 123	1.6643	0

120°	Nat.	Log.	Dif.	Nat.	Log.	Dif.	Nat.	Log.	Dif.	Log.	Nat.	59°
		Cosines.			Sines.			Cotangents.			Tangents.	

31°	Sines			Cosines			Tangents			Cotangents		148°
	Nat.	Log.	Dif.	Nat.	Log.	Dif.	Nat.	Log.	Dif.	Log.	Nat.	
0′	.51 504	9.71 184	.35	.85 717	9.93 307	.13	.60 086	9.77 877	.48	0.22 123	1.6648	60′
1	529	205		702	299		126	906		094	682	59
2	554	226		687	291	.12	165	935	.47	065	621	58
3	579	247		672	284	.13	205	963	.48	037	610	57
4	604	268		657	276	.12	245	992	.47	008	599	56
5	628	289		642	269	.13	284	9.78 020	.48	0.21 980	588	55
6	653	310		627	261		324	049	.47	951	577	54
7	678	331		612	253	.12	364	077	.48	923	566	53
8	708	352		507	246	.13	403	106		894	555	52
9	728	373	.33	582	238		443	135	.47	865	545	51
10	.51 753	9.71 393	.35	.85 567	9.93 230	.12	.60 483	9.78 163	.48	0.21 837	1.6584	50
11	778	414		551	223	.13	522	192	.47	808	523	49
12	808	435		586	215		562	220	.48	780	512	48
13	828	456		521	207	.12	602	249	.47	751	501	47
14	852	477		506	200	.13	642	.277	.48	723	490	46
15	877	498		491	192		681	306	.47	694	479	45
16	902	519	.38	476	184	.12	721	334	.48	666	469	44
17	927	539	.35	461	177	.13	761	363	.47	637	458	43
18	952	560		446	169		801	391		609	447	42
19	977	581		431	161	.12	841	419	.48	581	436	41
20	.52 002	9.71 602	.38	.85 416	9.93 154	.13	.60 881	9.78 448	.47	0.21 552	1.6426	40
21	026	622	.35	401	146		921	476	.48	524	415	39
22	051	643		385	138	.12	960	505	.47	495	404	38
23	076	664		370	131	.13	.61 000	533	.48	467	393	37
24	101	685	.38	355	123		040	562	.47	438	383	36
25	126	705	.35	340	115	.12	080	590		410	372	35
26	151	726		325	108	.13	120	618	.48	382	361	34
27	175	747	.38	310	100		160	647	.47	353	351	33
28	200	767	.85	294	092		200	675	.48	325	340	32
29	225	788		279	084	.12	240	704	.47	296	329	31
30	.52 250	9.71 809	.38	.85 264	9.93 077	.13	.61 280	9.78 732	.47	0.21 268	1.6319	30
31	275	829	.85	249	069		320	760	.48	240	308	29
32	299	850	.38	234	061		360	789	.47	211	297	28
33	824	870	.85	218	053	.12	400	817		183	287	27
34	849	891	.38	203	046	.13	440	845	.48	155	276	26
35	874	911	.85	188	038		480	874	.47	126	265	25
36	899	932	.38	173	030		520	902		098	255	24
37	423	952	.85	157	022		561	930	.48	070	244	23
38	448	973		142	014	.12	601	959	.47	041	234	22
39	473	994	.38	127	007	.13	641	987		013	224	21
40	.52 498	9.72 014	.38	.85 112	9.92 999	.13	.61 681	9.79 015	.47	0.20 985	1.6212	20
41	522	034	.35	•096	991		721	043	.48	957	202	19
42	547	055	.38	081	983	.12	761	072	.47	928	191	18
43	572	075	.35	066	976	.13	601	100		900	181	17
44	597	096	.38	051	968		542	128		872	170	16
45	621	116	.35	085	960		882	156	.48	844	160	15
46	646	137	.38	020	952		922	185	.47	815	149	14
47	671	157		005	944		962	213		787	139	13
48	696	177	.35	.84 989	936	.12	.62 006	241		759	128	12
49	720	198	.33	974	929	.13	043	269		731	118	11
50	.52 745	9.72 218	.33	.84 959	9.92 921	.13	.62 083	9.79 297	.48	0.20 703	1.6107	10
51	770	238	.35	943	913		124	326	.47	674	097	9
52	794	259	.38	928	905		164	354		646	087	8
53	819	279		913	897		204	382		618	076	7
54	844	299	.35	897	889		245	410		590	066	6
55	869	320	.33	882	881	.12	285	438		‘562	055	5
56	898	340		866	874	.13	325	466	.48	534	045	4
57	918	360	.35	851	866		366	495	.47	505	034	3
58	943	381	.33	836	858		406	523		477	024	2
59	967	401		820	850		446	551		449	014	1
60	.52 992	9.72 421		.84 605	9.92 842		.62 487	9.79 579		0.20 421	1.6003	0

121°	Nat.	Log.	Dif.	Nat.	Log.	Dif.	Nat.	Log.	Dif.	Log.	Nat.	58°
		Cosines.			Sines.			Cotangents.		Tangents.		

32°	SINES.			COSINES.			TANGENTS.			COTANGENTS.		147°
	Nat.	Log.	Dif.	Nat.	Log.	Dif.	Nat.	Log.	Dif.	Log.	Nat.	
0'	.52 992	9.72 421	.33	.84 805	9.92 842	.18	.62 487	9.79 579	.47	0.20 421	1.6008	60'
1	.53 017	441		789	834		527	607		393	1.5993	59
2	041	461	.35	774	826		568	635		365	983	58
3	066	482	.33	759	818		608	663		337	972	57
4	091	502		743	810	.12	649	691		309	962	56
5	115	522		729	803	.13	689	719		281	952	55
6	140	542		712	795		730	747	.48	253	941	54
7	164	562		697	787		770	776	.47	224	931	53
8	189	582		681	779		811	804		196	921	52
9	214	602		666	771		852	832		168	911	51
10	.53 238	9.72 622	.35	.84 650	9.92 763	.18	.62 892	9.79 860	.47	0.20 140	1.5900	50
11	263	643	.33	635	755		933	888		112	890	49
12	288	663		619	747		973	916		084	880	48
13	312	683		604	739		.63 014	944		056	869	47
14	337	703		588	731		055	972		028	859	46
15	361	723		578	723		095	9.80 000		000	849	45
16	386	743		557	715		136	028		0.19 972	839	44
17	411	763		542	707		177	056		944	829	43
18	435	783		526	699		217	084		916	818	42
19	460	803		511	691		258	112		888	808	41
20	.53 484	9.72 823	.33	.84 495	9.92 683	.18	.63 299	9.80 140	.47	0.19 860	1.5798	40
21	509	843		480	675		340	168	.45	832	788	39
22	534	863		464	667		380	195	.47	805	778	38
23	558	883	.32	449	659		421	223		777	769	37
24	583	902	.33	433	651		462	251		749	757	36
25	607	922		417	643		503	279		721	747	35
26	632	942		402	635		544	307		693	737	34
27	656	962		386	627		584	335		665	727	33
28	681	982		370	619		625	363		637	717	32
29	705	9.73 002		355	611		666	391		609	707	31
30	.53 730	9.73 022	.32	.84 339	9.92 603	.13	.63 707	9.80 419	.47	0.19 581	1.5697	30
31	754	041	.33	324	595		748	447	.45	553	687	29
32	779	061		308	587		789	474	.47	526	677	28
33	804	081		292	579		830	502		498	667	27
34	828	101		277	571		871	530		470	657	26
35	853	121	.32	261	563		912	558		442	647	25
36	877	140	.33	245	555	.15	953	586		414	637	24
37	902	160		230	546	.13	994	614		386	627	23
38	926	180		214	538		.64 035	642	.45	358	617	22
39	951	200	.32	198	530		076	669	.47	331	607	21
40	.53 975	9.73 219	.33	.84 182	9.92 522	.18	.64 117	9.80 697	.47	0.19 303	1.5597	20
41	.54 000	239		167	514		158	725		275	587	19
42	024	259	.32	151	506		199	753		247	577	18
43	049	278	.33	135	498		240	781	.45	219	567	17
44	073	298		120	490		281	808	.47	192	557	16
45	097	318	.32	104	482	.15	322	836		164	547	15
46	122	337	.33	088	473	.13	363	864		136	537	14
47	146	357		072	465		404	892	.45	108	527	13
48	171	377	.32	057	457		446	919	.47	081	517	12
49	195	396	.33	041	449		487	947		053	507	11
50	.54 220	9.73 416	.32	.84 025	9.92 441	.18	.64 528	9.80 975	.47	0.19 025	1.5497	10
51	244	435	.38	009	433		569	9.81 003	.45	0.18 997	487	9
52	269	455	.32	.83 994	425	.15	610	030	.47	970	477	8
53	293	474	.33	978	416	.13	652	058		942	468	7
54	317	494	.32	962	408		693	086	.45	914	458	6
55	342	513	.33	946	400		734	113	.47	887	448	5
56	366	533	.32	930	392		775	141		859	438	4
57	391	552	.33	915	384		817	169	.45	831	428	3
58	415	572	.32	899	376	.15	858	196	.47	804	418	2
59	440	591	.33	883	367	.13	899	224		776	408	1
60	.54 464	9.73 611		.83 867	9.92 359		.64 941	9.81 252		0.18 748	1.5399	0

122°	Nat.	Log.	Dif.	Nat.	Log.	Dif.	Nat.	Log.	Dif.	Log.	Nat.	57°
		COSINES.			SINES.			COTANGENTS.			TANGENTS.	

33°	SINES.			COSINES.			TANGENTS.			COTANGENTS.		146°
	Nat.	Log.	Dif.	Nat.	Log.	Dif.	Nat.	Log.	Dif.	Log.	Nat.	
0′	.54 464	9.73 611	.32	.83 867	9.92 359	.18	.64 941	9.81 252	.45	0.18 748	1.5399	60′
1	488	630	.33	851	351		982	279	.47	721	389	59
2	518	650	.32	835	343		.65 024	307		693	379	58
3	587	669	.33	819	335	.15	065	335	.45	665	369	57
4	561	689	.82	804	326	.13	106	362	.47	638	859	56
5	586	708		788	318		148	390		610	850	55
6	610	727	.33	772	310		189	418	.45	582	840	54
7	635	747	.32	756	302	.15	281	445	.47	555	880	53
8	659	766		740	293	.13	272	473	.45	527	820	52
9	683	785	.33	724	285		814	500	.47	500	811	51
10	.54 708	9.73 805	.32	.83 708	9.92 277	.13	.65 855	9.81 528	.47	0.18 472	1.5801	50
11	732	824		692	269	.15	897	556	.45	444	291	49
12	756	843	.33	676	260	.13	438	583	.47	417	282	48
13	781	863	.32	660	252		480	611	.45	389	272	47
14	805	882		645	244	.15	521	638	.47	362	262	46
15	829	901	.33	629	235	.13	563	666	.45	334	258	45
16	854	921	.32	613	227		604	693	.47	307	243	44
17	878	940		597	219		646	721	.45	279	238	43
18	902	959		581	211	.15	688	748	.47	252	224	42
19	927	978		565	202	.13	729	776	.45	224	214	41
20	54 951	9.73 997	.33	.83 549	9.92 194	.13	.65 771	9.81 803	.47	0.18 197	1.5204	40
21	975	9.74 017	.32	533	186	.15	813	831	.45	169	195	39
22	999	036		517	177	.13	854	858	.47	142	185	38
23	.55 024	055		501	169		896	886	.45	114	175	37
24	048	074		485	161	.15	938	913	.47	087	166	36
25	072	093	.33	469	152	.13	960	941	.45	059	156	35
26	097	113	.32	453	144		.66 021	968	.47	032	147	34
27	121	132		437	136	.15	063	996	.45	004	137	33
28	145	151		421	127	.13	105	9.82 023	.47	0.17 977	127	32
29	169	170		405	119		147	051	.45	949	118	31
30	.55 194	9.74 189	.32	.83 889	9.92 111	.15	.66 189	9.82 078	.47	0.17 922	1.5108	30
31	218	208		373	102	.13	230	106	.45	894	099	29
32	242	227		356	094		272	133	.47	867	089	28
33	266	246		340	086	.15	314	161	.45	839	080	27
34	291	265		324	077	.13	356	188		812	070	26
35	315	284		308	069	.15	398	215	.47	785	061	25
36	339	303		292	060	.13	440	243	.45	757	051	24
37	363	322		276	052		482	270	.47	730	042	23
38	388	341		260	044	.15	524	298	.45	702	032	22
39	412	360		244	035	.13	566	325		675	023	21
40	.55 436	9.74 379	.32	.83 228	9.92 027	.15	.66 608	9.82 352	.47	0.17 648	1.5013	20
41	460	398		212	018	.13	650	380	.45	620	004	19
42	484	417		195	010		692	407	.47	593	1.4994	18
43	509	436		179	002	.15	784	435	.45	565	985	17
44	533	455		163	9.91 993	.13	776	462		538	975	16
45	557	474		147	985	.15	818	489	.47	511	966	15
46	581	493		131	976	.13	860	517	.45	483	957	14
47	605	512		115	968	.15	902	544		456	947	13
48	630	531	.30	096	959	.13	944	571	.47	429	938	12
49	654	549	.32	082	951	.15	986	599	.45	401	928	11
50	.55 678	9.74 568	.32	.83 066	9.91 942	.13	.67 028	9.82 626	.45	0.17 374	1.4919	10
51	702	587		050	934	.15	071	653	.47	347	910	9
52	726	606		034	925	.13	113	681	.45	319	900	8
53	750	625		017	917	.15	155	708		292	891	7
54	775	644	.30	001	908	.13	197	735		265	882	6
55	799	662	.32	.92 985	900	.15	239	762	.47	238	872	5
56	823	681		969	891	.13	282	790	.45	210	863	4
57	847	700		958	883	.15	324	817		183	654	3
58	871	719	.30	936	874	.13	366	844		156	844	2
59	895	737	.32	920	866	.15	409	871	.47	129	835	1
60	.55 919	9.74 756		.52 904	9.91 857		.67 451	9.82 899		0.17 101	1.4826	0

123°	Nat.	Log.	Dif.	Nat.	Log.	Dif.	Nat.	Log.	Dif.	Log.	Nat.	56°
		COSINES.			SINES.			COTANGENTS.			TANGENTS.	

34°	Sines.			Cosines.			Tangents.			Cotangents.		145°
	Nat.	Log.	Dif.	Nat.	Log.	Dif.	Nat.	Log.	Dif.	Log.	Nat.	
0′	.55 919	9.74 756	.32	.82 904	9.91 857	.18	.67 451	9.82 899	.45	0.17 101	1.4826	60′
1	943	775		887	849	.15	493	926		074	816	59
2	968	794	.30	871	840	.18	586	953		047	807	58
3	992	812	.32	855	832	.15	578	980	.47	020	798	57
4	.56 016	831		839	823	.18	620	9.83 008	.45	0.16 992	788	56
5	040	850	.30	822	815	.15	663	035		965	779	55
6	064	868	.32	806	806	.18	705	062		938	770	54
7	088	887		790	798	.15	748	089	.47	911	761	53
8	112	906	.30	773	789	.18	790	117	.45	883	751	52
9	136	924	.32	757	781	.15	832	144		856	742	51
10	.56 160	9.74 943	.30	.82 741	9.91 772	.15	.67 875	9.83 171	.45	0.16 829	1.4733	50
11	184	961	.32	724	763	.18	917	198		802	724	49
12	208	980		708	755	.15	960	225		775	715	48
13	232	999	.30	692	746	.18	.68 002	252	.47	748	705	47
14	256	9.75 017	.32	675	738	.15	045	280	.45	720	696	46
15	280	036	.30	659	729		088	307		693	687	45
16	305	054	.32	643	720	.18	130	334		666	678	44
17	329	073	.30	626	712	.15	173	361		639	669	43
18	353	091	.32	610	703	.18	215	388		612	659	42
19	377	110	.30	593	695	.15	258	415		585	650	41
20	.56 401	9.75 128	.32	.82 577	9.91 686	.15	.68 301	9.83 442	.47	0.16 558	1.4641	40
21	425	147	.30	561	677	.18	343	470	.45	530	632	39
22	449	165	.32	544	669	.15	386	497		503	623	38
23	473	184	.30	528	660		429	524		476	614	37
24	497	202	.32	511	651	.18	471	551		449	605	36
25	521	221	.30	495	643	.15	514	578		422	596	35
26	545	239	.32	478	634		557	605		395	586	34
27	569	258	.30	462	625	.18	600	632		368	577	33
28	593	276		446	617	.15	642	659		341	568	32
29	617	294	.32	429	608		685	686		314	559	31
30	.56 641	9.75 313	.30	.82 413	9.91 599	.18	.68 728	9.83 713	.45	0.16 287	1.4550	30
31	665	331	.32	396	591	.15	771	740	.47	260	541	29
32	689	350	.30	380	582		814	768	.45	232	532	28
33	713	368		363	573	.18	857	795	.	205	523	27
34	738	386	.32	347	565	.15	900	822		178	514	26
35	760	405	.30	330	556		942	849		151	505	25
36	784	423		314	547		985	876		124	496	24
37	809	441		297	538	.18	.69 028	903		097	487	23
38	832	459	.32	281	530	.15	071	930		070	478	22
39	855	478	.30	264	521		114	957		043	469	21
40	.56 880	9.75 496	.30	.82 248	9.91 512	.18	.69 157	9.83 984	.45	0.16 016	1.4460	20
41	904	514	.32	231	504	.15	200	9.84 011		0.15 989	451	19
42	928	533	.30	214	495		243	038		962	442	18
43	952	551		198	486	.	286	065		935	433	17
44	976	569		181	477	.18	329	092		908	424	16
45	.57 000	587		165	469	.15	372	119		881	415	15
46	024	605	.32	148	460		416	146		854	406	14
47	047	624	.30	132	451		459	173		827	397	13
48	071	642		115	442		502	200		800	388	12
49	095	660		098	433	.18	545	227		773	379	11
50	.57 119	9.75 678	.30	.82 082	9.91 425	.15	.69 588	9.84 254	.43	0.15 746	1.4370	10
51	143	696		065	416		631	280	.45	720	361	9
52	167	714	.32	049	407		675	307		693	352	8
53	191	733	.30	032	398		718	334		666	344	7
54	215	751		015	389	.18	761	361		639	335	6
55	238	769		.81 999	381	.15	804	388		612	326	5
56	262	787		982	372		847	415		585	317	4
57	286	805		965	363		891	442		558	308	3
58	310	823		949	354		934	469		531	299	2
59	334	841		932	345		977	496		504	290	1
60	.57 358	9.75 859		.81 915	9.91 336		.70 021	9.84 523		0.15 477	1.4281	0
124°	Nat.	Log.	Dif.	Nat.	Log.	Dif.	Nat.	Log.	Dif.	Log.	Nat.	55°
		Cosines.			Sines.			Cotangents.			Tangents.	

35°	SINES.			COSINES.			TANGENTS.			COTANGENTS.		144°
	Nat.	Log.	Dif.	Nat.	Log.	Dif.	Nat.	Log.	Dif.	Log.	Nat.	
0'	.57 358	9.75 859	.80	.81 915	9.91 336	.13	.70 021	9.84 523	.45	0.15 477	1.4281	60'
1	381	877		899	328	.15	064	550	.43	450	273	59
2	405	895		882	319		107	576	.45	424	264	58
3	429	913		865	310		151	603		397	255	57
4	453	931		848	301		194	630		370	246	56
5	477	949		832	292		238	657		343	237	55
6	501	967		815	283		281	684		316	229	54
7	524	985		798	274	.18	325	711		289	220	53
8	548	9.76 003		782	266	.15	368	738	.43	262	211	52
9	572	021		765	257		412	764	.45	236	202	51
10	.57 596	9.76 039	.80	.81 748	9.91 248	.15	.70 455	9.84 791	.45	0.15 209	1.4193	50
11	619	057		731	239		499	818		182	185	49
12	643	075		714	230		542	845		155	176	48
13	667	093		698	221		586	872		128	167	47
14	691	111		681	212		629	899	.43	101	158	46
15	715	129	.28	664	203		673	925	.45	075	150	45
16	738	146	.30	647	194		717	952		048	141	44
17	762	164		631	185		760	979		021	132	43
18	786	182		614	176		804	9.85 006		0.14 994	124	42
19	810	200		597	167		848	033	.43	967	115	41
20	.57 833	9.76 218	.80	.81 580	9.91 158	.15	.70 891	9.85 059	.45	0.14 941	1.4106	40
21	857	236	.28	563	149	.18	935	086		914	097	39
22	881	253	.30	546	141	.15	979	113		887	069	38
23	904	271		530	132		.71 023	140	.43	860	080	37
24	928	289		513	123		066	166	.45	834	071	36
25	952	307	.28	496	114		110	193		807	063	35
26	976	324	.30	479	105		154	220		780	054	34
27	999	342		462	096		198	247	.43	753	045	33
28	.58 023	360		445	087		242	273	.45	727	037	32
29	047	378	.28	428	078		285	300		700	028	31
30	.58 070	9.76 395	.80	.81 412	9.91 069	.15	.71 329	9.85 327	.45	0.14 673	1.4019	30
31	094	413		395	060		373	354	.43	646	011	29
32	118	431	.28	378	051		417	380	.45	620	002	28
33	141	448	.30	361	042		461	407		593	1.3994	27
34	165	466		344	033	.17	505	434	.43	566	985	26
35	189	484	.28	327	023	.15	549	460	.45	540	976	25
36	212	501	.30	310	014		593	487		513	968	24
37	236	519		293	005		637	514	.43	486	959	23
38	260	537	.28	276	9.90 996		681	540	.45	460	951	22
39	283	554	.30	259	987		725	567		433	942	21
40	.58 307	9.76 572	.30	.81 242	9.90 978	.15	.71 769	9.85 594	.43	0.14 406	1.3934	20
41	330	590	.28	225	969		813	620	.45	380	925	19
42	354	607	.80	208	960		857	647		353	916	18
43	378	625	.28	191	951		901	674	.43	326	908	17
44	401	642	.30	174	942		946	700	.45	300	899	16
45	425	660	.28	157	933		990	727		273	891	15
46	449	677	.30	140	924		.72 034	754	.43	246	882	14
47	472	695	.29	123	915		078	780	.45	220	874	13
48	496	712	.30	106	906	.17	122	807		193	865	12
49	519	730	.28	089	896	.15	167	834	.43	166	857	11
50	.58 543	9.76 747	.30	.81 072	9.90 887	.15	.72 211	9.85 860	.45	0.14 140	1.3848	10
51	567	765	.28	055	878		255	887	.43	113	840	9
52	590	782	.30	038	869		299	913	.45	087	831	8
53	614	800	.28	021	860		344	940		060	823	7
54	637	817	.30	004	851		388	967	.43	033	814	6
55	661	835	.28	.80 987	842	.17	432	993	.45	007	806	5
56	684	852	.30	970	832	.15	477	9.86 020	.43	0.13 980	798	4
57	708	870	.28	953	823		521	046	.45	954	789	3
58	731	887		936	814		565	073		927	781	2
59	755	904	.80	919	805		610	100	.43	900	772	1
60	.58 779	9.76 922		.80 902	9.90 796		.72 654	9.86 126		· 0.13 874	1.3764	0
125°	Nat.	Log.	Dif.	Nat.	Log.	Dif.	Nat.	Log.	Dif.	Log.	Nat.	54°
		COSINES.			SINES.			COTANGENTS.			TANGENTS.	

36°	Sines.			Cosines.			Tangents.			Cotangents.		143°
	Nat.	Log.	Dif.	Nat.	Log.	Dif.	Nat.	Log.	Dif.	Log.	Nat.	
0'	.58 779	9.76 922	.28	.80 902	9.90 796	.15	.72 654	9.86 126	.45	0.13 874	1.3764	60'
1	802	939	.30	885	787	.17	699	153	.43	847	755	59
2	826	957	.28	667	777	.15	743	179	.45	821	747	58
3	849	974		850	768		788	206	.43	794	739	57
4	873	991	.30	833	759		832	232	.45	768	730	56
5	896	9.77 009	.28	816	750		877	259	.43	741	722	55
6	920	026		799	741	.17	921	285	.45	715	713	54
7	943	043	.30	782	731	.15	966	312	.43	688	705	53
8	967	061	.28	765	722		.73 010	338	.45	662	697	52
9	990	078		748	713		055	365		635	688	51
10	.59 014	9.77 095	.28	.80 730	9.90 704	.17	.73 100	9.86 392	.43	0.13 608	1.3680	50
11	037	112	.30	713	694	.15	144	418	.45	582	672	49
12	061	130	.28	696	685		189	445	.43	555	663	48
13	084	147		679	676		234	471	.45	529	655	47
14	108	164		662	667	.17	278	498	.43	502	647	46
15	131	181	.30	644	657	.15	323	524	.45	476	638	45
16	154	199	.28	627	648		368	551	.43	449	630	44
17	178	216		610	639		413	577		423	622	43
18	201	233		593	630	.17	457	603	.45	397	613	42
19	225	250	.30	576	620	.15	502	630	.43	370	605	41
20	.59 248	9.77 268	.28	.80 558	9.90 611	.15	.73 547	9.86 656	.45	0.13 344	1.3597	40
21	272	285		541	602	.17	592	683	.43	317	588	39
22	295	302		524	592	.15	637	709	.45	291	580	38
23	318	319		507	583		681	736	.43	264	572	37
24	342	336		489	574		726	762	.45	238	564	36
25	365	353		472	565	.17	771	789	.43	211	555	35
26	389	370		455	555	.15	816	815	.45	185	547	34
27	412	387	.30	438	546		861	842	.43	158	539	33
28	436	405	.28	420	537	.17	906	868		132	531	32
29	459	422		403	527	.15	951	894	.45	106	522	31
30	.59 462	9.77 439	.28	.80 386	9.90 518	.15	.73 996	9.86 921	.43	0.13 079	1.3514	30
31	506	456		368	509	.17	.74 041	947	.45	053	506	29
32	529	473		351	499	.15	086	974	.43	026	498	28
33	552	490		334	490	.17	131	9.87 000	.45	000	490	27
34	576	507		316	480	.15	176	027	.43	0.12 973	481	26
35	599	524		299	471		221	053		947	473	25
36	622	541		282	462	.17	267	079	.45	921	465	24
37	646	558		264	452	.15	312	106	.43	894	457	23
38	669	575		247	443		357	132		868	449	22
39	693	592		230	434	.17	402	158	.45	842	440	21
40	.59 716	9.77 609	.28	.80 212	9.90 424	.15	.74 447	9.87 185	.43	0.12 815	1.3432	20
41	739	626		195	415	.17	492	211	.45	789	424	19
42	768	643		178	405	.15	538	238	.43	762	416	18
43	786	660		160	396	.17	583	264		736	408	17
44	809	677		143	386	.15	628	290	.45	710	400	16
45	832	694		125	377		674	317	.43	683	392	15
46	856	711		108	368	.17	719	343		657	384	14
47	879	728	.27	091	358	.15	764	369	.45	631	375	13
48	902	744	.28	073	349	.17	810	396	.43	604	367	12
49	926	761		056	339	.15	855	422		578	359	11
50	.59 949	9.77 778	.28	.80 038	9.90 330	.17	.74 900	9.87 448	.45	0.12 552	1.3351	10
51	972	795		021	320	.15	946	475	.43	525	343	9
52	995	812		003	311	.17	991	501		499	335	8
53	.60 019	829		.79 986	301	.15	.75 037	527	.45	473	327	7
54	042	846	.27	968	292	.17	082	554	.43	446	319	6
55	065	862	.28	951	282	.15	128	580		420	311	5
56	089	879		934	273	.17	173	606	.45	394	303	4
57	112	896		916	263	.15	219	633	.43	367	295	3
58	135	913		899	254	.17	264	659		341	287	2
59	158	930	.27	881	244	.15	310	685		315	278	1
60	.60 182	9.77 946		.79 864	9.90 235		.75 355	9.87 711		0.12 289	1.3270	0

126°	Nat.	Log.	Dif.	Nat.	Log.	Dif.	Nat.	Log.	Dif.	Log.	Nat.	53°
		Cosines.			Sines.			Cotangents.			Tangents.	

37°	Sines.			Cosines.			Tangents.			Cotangents.		142°
	Nat.	Log.	Dif.	Nat.	Log.	Dif.	Nat.	Log.	Dif.	Log.	Nat.	
0'	.60 182	9.77 946	.28	.79 864	9.90 235	.17	.75 355	9.87 711	.45	0.12 289	1.3270	60'
1	205	963		846	225	.15	401	738	.43	262	262	59
2	228	980		829	216	.17	447	764		236	254	58
3	251	997	.27	811	206	.15	492	790	.45	210	246	57
4	274	9.78 013	.28	793	197	.17	588	817	.43	183	238	56
5	298	030		776	187	.15	584	843		157	230	55
6	321	047	.27	758	178	.17	629	869		131	222	54
7	344	063	.28	741	168	.15	675	895	.45	105	214	53
8	367	080		723	159	.17	721	922	.43	078	206	52
9	390	097	.27	706	149		767	948		052	196	51
10	.60 414	9.78 113	.28	.79 688	9.90 139	.15	.75 812	9.87 974	.49	0.12 026	1.3190	50
11	437	130		671	130	.17	858	9.88 000	.45	000	182	49
12	460	147	.27	653	120	.15	904	027	.43	0.11 973	175	48
13	483	163	.28	635	111	.17	950	053		947	167	47
14	506	180		618	101		996	079		921	159	46
15	520	197	.27	600	091	.15	.76 042	105		895	151	45
16	553	213	.28	563	082	.17	088	131	.45	869	143	44
17	576	230	.27	565	072	.15	134	158	.43	842	135	43
18	599	246	.28	547	063	.17	180	184		816	127	42
19	622	263		530	053		226	210		790	119	41
20	.60 645	9.78 280	.27	.79 512	9.90 043	.15	.76 272	9.88 236	.43	0.11 764	1.3111	40
21	666	296	.28	494	034	.17	318	262	.45	738	103	39
22	691	313	.27	477	024		364	289	.43	711	095	38
23	714	329	.26	459	014	.15	410	315		685	087	37
24	738	346	.27	441	005	.17	456	341		659	079	36
25	761	362	.28	424	9.89 995		502	367		633	072	35
26	784	379	.27	406	985	.15	548	393	.45	607	064	34
27	807	395	.28	388	976	.17	594	420	.43	580	056	33
28	830	412	.27	371	966		640	446		554	048	32
29	853	428	.28	353	956	.15	686	472		528	040	31
30	.60 876	9.78 445	.27	.79 335	9.89 947	.17	.76 788	9.88 498	.43	0.11 502	1.3032	30
31	899	461	.28	318	937		779	524		476	024	29
32	922	478	.27	300	927	.15	825	550	.45	450	017	28
33	945	494		282	918	.17	871	577	.43	423	009	27
34	968	510	.28	264	908		918.	603		397	001	26
35	991	527	.27	247	898		964	629		371	1.2993	25
36	.61 015	543	.28	229	888	.15	.77 010	655		345	985	24
37	038	560	.27	211	879	.17	057	681		319	977	23
38	061	576		193	869		103	707		293	970	22
39	084	592	.28	176	859		149	733		267	962	21
40	.61 107	9.78 609	.27	.79 158	9.89 849	.15	.77 196	9.88 759	.45	0.11 241	1.2954	20
41	130	625	.28	140	840	.17	242	786	.43	214	946	19
42	153	642	.27	122	830		289	812		188	938	18
43	176	658		105	820		335	838		162	931	17
44	199	674	.28	087	810	.15	382	864		136	923	16
45	222	691	.27	069	801	.17	428	890		110	915	15
46	245	707		051	791		475	916		084	907	14
47	268	723		033	781		521	942		058	900	13
48	291	739	.28	016	771		568	968		032	892	12
49	314	756	.27	.78 998	761	.15	615	994		006	884	11
50	.61 337	9.78 772	.27	.78 980	9.89 752	.17	.77 661	9.89 020	.43	0.10 980	1.2876	10
51	360	788	.28	962	742		708	046	.45	954	869	9
52	383	805	.27	944	732		754	073	.43	927	861	8
53	406	821		926	722		801	099		901	853	7
54	429	837		908	712		848	125		875	846	6
55	451	853		891	702	.15	895	151		849	838	5
56	474	869	.28	873	693	.17	941	177		823	830	4
57	497	886	.27	855	683		988	203		797	822	3
58	520	902		837	673	.	.78 035	229		771	815	2
59	543	918		819	663		082	255		745	807	1
60	.61 566	9.78 934		.78 801	9.89 653		.78 129	9.89 281		0.10 719	1.2799	0

38°	Sines.			Cosines.			Tangents.			Cotangents.		141°
	Nat.	Log.	Dif.	Nat.	Log.	Dif.	Nat.	Log.	Dif.	Log.	Nat.	
0′	.61 566	9.78 934	.27	.78 801	9.89 653	.17	.78 129	9.89 281	.43	0.10 719	1.2799	60′
1	589	950	.28	783	643		175	307		693	792	59
2	612	967	.27	765	633	.15	222	333		667	784	58
3	635	983		747	624	.17	269	359		641	776	57
4	658	999		729	614		316	385		615	769	56
5	681	9.79 015		711	604		363	411		589	761	55
6	704	031	.	694	594		410	437		563	753	54
7	726	047		676	584		457	463		537	746	53
8	749	063		658	574		504	489		511	738	52
9	772	079		640	564		551	515		485	731	51
10	.61 795	9.79 095	.27	.78 622	9.89 554	.17	.78 598	9.89 541	.43	0.10 459	1.2723	50
11	818	111	.28	604	544		645	567		433	715	49
12	841	128	.27	586	534		692	593		407	708	48
13	864	144		568	524		739	619		381	700	47
14	887	160		550	514		786	645		355	693	46
15	909	176		532	504	.15	834	671		329	685	45
16	932	192		514	495	.17	881	697		303	677	44
17	955	208		496	485		928	723		277	670	43
18	978	224		478	475		975	749		251	662	42
19	.62 001	240		460	465		.79 022	775		225	655	41
20	.62 024	9.79 256	.27	.78 442	9.89 455	.17	.79 070	9.89 801	.43	0.10 199	1.2647	40
21	046	272		424	445		117	827		173	640	39
22	069	288		405	435		164	853		147	632	38
23	092	304	.25	387	425		212	879		121	624	37
24	115	319	.27	369	415	.	259	905		095	617	36
25	138	335		351	405		306	931		069	609	35
26	160	351		333	395		354	957		043	602	34
27	183	367		315	385		401	983		017	594	33
28	206	383		297	375	.18	449	9.90 009		0.09 991	587	32
29	229	399		279	364	.17	496	035		965	579	31
30	.62 251	9.79 415	.27	.78 261	9.89 354	.17	.79 544	9.90 061	.42	0.09 939	1.2572	30
31	274	431		243	344		591	086	.43	914	564	29
32	297	447		225	334		639	112		888	557	28
33	320	463	.25	206	324		686	138		862	549	27
34	342	478	.27	188	314		734	164		836	542	26
35	365	494		170	304		781	190		810	534	25
36	388	510		152	294		829	216		784	527	24
37	411	526		134	284		877	242		758	519	23
38	433	542		116	274		924	268		732	512	22
39	456	558	.25	098	264		972	294		706	504	21
40	.62 479	9.79 573	.27	.78 079	9.89 254	.17	.80 020	9.90 320	.43	0.09 680	1.2497	20
41	502	589		061	244	.18	067	346	.42	654	489	19
42	524	605		043	233	.17	115	371	.43	629	482	18
43	547	621	.25	025	223		163	397		603	475	17
44	570	636	.27	007	213		211	423		577	467	16
45	592	652		.77 988	203		258	449		551	460	15
46	615	668		970	193		306	475		525	452	14
47	638	684	.25	952	183		354	501		499	445	13
48	660	699	.27	934	173	.18	402	527		473	437	12
49	683	715		916	162	.17	450	553	.42	447	430	11
50	.62 706	9.79 731	.25	.77 897	9.89 152	.17	.80 498	9.90 578	.43	0.09 422	1.2423	10
51	728	746	.27	879	142		546	604		396	415	9
52	751	762		861	132		594	630		370	408	8
53	774	778	.25	843	122		642	656		344	401	7
54	796	793	.27	824	112	.18	690	682		318	393	6
55	819	809		806	101	.17	738	708		292	386	5
56	842	825	.25	788	091		786	734	.42	266	378	4
57	864	840	.27	769	081		834	759	.43	241	371	3
58	887	856		751	071	.18	882	785		215	364	2
59	909	872	.25	733	060	.17	930	811		189	356	1
60	.62 932	9.79 887		.77 715	9.89 050		.80 978	9.90 837		0.09 163	1.2349	0

128°	Nat.	Log.	Dif.	Nat.	Log.	Dif.	Nat.	Log.	Dif.	Log.	Nat.	51°
		Cosines.			Sines.			Cotangents.			Tangents.	

39°	SINES			COSINES			TANGENTS			COTANGENTS		140°
	Nat.	Log.	Dif.	Nat.	Log.	Dif.	Nat.	Log.	Dif.	Log.	Nat.	
0′	.62 932	9.79 887	.27	.77 715	9.89 050	.17	.80 978	9.90 837	.43	0.09 163	1.2349	60′
1	955	903	.25	696	040		.81 027	863		137	842	59
2	977	918	.27	678	030		075	889	.42	111	834	58
3	.63 000	934		660	020	.18	128	914	.43	086	327	57
4	022	950	.25	641	009	.17	171	940		060	820	56
5	045	965	.27	623	8.88 999		220	966		034	312	55
6	068	981	.25	605	989	.18	268	992		008	805	54
7	090	996	.27	586	978	.17	316	9.91 018	.42	0.08 982	298	53
8	118	9.80 012	.25	568	968		364	043	.43	957	290	52
9	185	027	.27	550	958		413	069		931	283	51
10	.63 158	9.80 043	.25	.77 531	9.88 948	.18	.81 461	9.91 095	.43	0.08 905	1.2276	50
11	180	058	.27	513	937	.17	510	121		879	268	49
12	203	074	.25	494	927		558	147	.42	853	261	48
13	225	089	.27	476	917	.18	606	172	.43	828	254	47
14	248	105	.25	456	906	.17	655	198		802	247	46
15	271	120	.27	439	896		703	224		776	239	45
16	293	136	.25	421	886	.18	752	250		750	232	44
17	316	151		402	875	.17	800	276	.42	724	225	43
18	338	166	.27	384	865		849	301	.43	699	218	42
19	361	182	.25	366	855	.18	898	327		673	210	41
20	.63 363	9.80 197	.27	.77 347	9.88 844	.17	.81 946	9.91 353	.43	0.08 647	1.2203	40
21	406	213	.25	329	834		995	379	.42	621	196	39
22	428	228	.27	310	824	.18	.82 044	404	.43	596	189	38
23	451	244	.25	292	813	.17	092	430		570	181	37
24	473	259		273	803		141	456		544	174	36
25	496	274	.27	255	793	.18	190	482	.42	518	167	35
26	518	290	.25	236	782	.17	238	507	.43	493	160	34
27	540	305		218	772	.18	287	533		467	153	33
28	563	320	.27	199	761	.17	336	559		441	145	32
29	585	336	.25	181	751		385	585	.42	415	138	31
30	.63 608	9.80 351	.25	.77 162	9.88 741	.18	.82 484	9.91 610	.43	0.08 390	1.2131	30
31	630	366	.27	144	730	.17	433	636		364	124	29
32	653	382	.25	125	720	.18	531	662		338	117	28
33	675	397		107	709	.17	580	688	.42	312	109	27
34	698	412	.27	088	699	.18	629	713	.43	287	102	26
35	720	428	.25	070	688	.17	678	739		261	095	25
36	742	443		051	678		727	765		235	088	24
37	765	458		033	668	.18	776	791	.42	209	081	23
38	787	473	.27	014	657	.17	825	816	.43	184	074	22
39	810	489	.25	.76 996	647	.18	874	842		158	066	21
40	.63 832	9.80 504	.25	.76 977	9.88 636	.17	.82 923	9.91 868	.42	0.08 132	1.2059	20
41	854	519		959	626	.18	972	893	.43	107	052	19
42	877	534	.27	940	615	.17	.83 022	919		081	045	18
43	899	550	.25	921	605	.18	071	945		055	038	17
44	922	565		903	594	.17	120	971	.42	029	031	16
45	944	580		884	584	.18	169	996	.43	004	024	15
46	966	595		866	573	.17	218	9.92 022		0.07 978	017	14
47	989	610		847	563	.18	268	048	.42	952	009	13
48	.64 011	625	.27	828	552	.17	317	073	.43	927	002	12
49	033	641	.25	810	542	.18	366	099		901	1.1995	11
50	.64 056	9.80 656	.25	.76 791	9.88 531	.17	.83 415	9.92 125	.42	0.07 875	1.1988	10
51	078	671		772	521	.18	465	150	.43	850	981	9
52	100	686		754	510		514	176		824	974	8
53	123	701		735	499	.17	564	202	.42	798	967	7
54	145	716		717	489	.18	613	227	.43	773	960	6
55	167	731		698	478	.17	662	253		747	953	5
56	190	746	.27	679	468	.18	712	279	.42	721	946	4
57	212	762	.25	661	457	.17	761	304	.43	696	939	3
58	234	777		642	447	.18	811	330		670	932	2
59	256	792		623	436		860	356	.42	644	925	1
60	.64 279	9.80 807		.76 604	9.88 425		.83 910	9.92 381		0.07 619	1.1918	0

129°	Nat.	Log.	Dif.	Nat.	Log.	Dif.	Nat.	Log.	Dif.	Log.	Nat.	50°
		COSINES.			SINES.			COTANGENTS.			TANGENTS.	

40°	Sines.			Cosines.			Tangents.			Cotangents.		139°
	Nat.	Log.	Dif.	Nat.	Log.	Dif.	Nat.	Log.	Dif.	Log.	Nat.	
0'	.64 279	9.80 807	.25	.76 604	9.88 425	.17	.83 910	9.92 381	.43	0.07 619	1.1918	60'
1	301	822		586	416	.18	960	407		593	910	59
2	323	837		567	404	.17	.84 009	433	.42	567	903	58
3	346	852		548	394	.16	059	458	.43	542	896	57
4	368	867		530	383		108	484		516	889	56
5	390	882		511	372	.17	158	510	.42	490	882	55
6	412	897		492	362	.18	208	535	.43	465	875	54
7	435	912		473	351		258	561		439	808	53
8	457	927		455	340	.17	307	587	.42	413	861	52
9	479	942		436	330	.18	357	612	.43	388	854	51
10	.64 501	9.80 957	.25	.76 417	9.88 319	.18	.84 407	9.92 638	.42	0.07 362	1.1847	50
11	524	972		398	308	.17	457	663	.43	337	840	49
12	546	987		380	298	.18	507	689		311	833	48
13	568	9.81 002		361	287		556	715	.42	285	826	47
14	590	017		342	276	.17	606	740	.43	260	819	46
15	612	032		323	266	.18	656	766		234	812	45
16	635	047	.23	304	255		706	792	.42	208	806	44
17	657	061	.25	286	244	.17	756	817	.43	183	799	43
18	679	076		267	234	.18	806	843	.42	157	792	42
19	701	091		248	223		856	868	.43	132	785	41
20	.64 723	9.81 106	.25	.76 229	9.88 212	.16	.84 906	9.92 894	.43	0.07 106	1.1778	40
21	746	121		210	201	.17	956	920	.42	080	771	39
22	768	136		192	191	.18	.85 006	945	.43	055	764	38
23	790	151		173	180		057	971	.42	029	757	37
24	812	166	.23	154	169		107	996	.43	004	750	36
25	834	180	.25	135	158	.17	157	9.93 022		0.06 978	743	35
26	856	195		116	148	.18	207	048	.42	952	736	34
27	878	210		097	137		257	073	.43	927	729	33
28	901	225		078	126		308	099	.42	901	722	32
29	923	240	.23	059	115	.17	358	124	.43	876	715	31
30	.64 945	9.81 254	.25	.76 041	9.88 105	.18	.85 408	9.93 150	.42	0.06 850	1.1708	30
31	967	269		022	094		458	175	.43	825	702	29
32	989	284		003	083		509	201		799	695	28
33	.65 011	299		.75 984	072		559	227	.42	773	689	27
34	033	314	.23	965	061	.17	609	252	.43	748	681	26
35	055	328	.25	946	051	.18	660	278	.42	722	674	25
36	077	343		927	040		710	303	.43	697	667	24
37	100	358	.23	908	029		761	329	.43	671	660	23
38	122	372	.25	889	018		811	354	.43	646	653	22
39	144	387		870	007		862	380		620	647	21
40	.65 166	9.81 402	.25	.75 851	9.87 996	.18	.85 912	9.93 406	.42	0.06 594	1.1640	20
41	188	417	.23	832	985	.17	963	431	.43	569	633	19
42	210	431	.25	813	975	.18	.86 014	457	.42	543	626	18
43	232	446		794	964		064	482	.43	518	619	17
44	254	461	.23	775	953		115	508	.42	492	612	16
45	276	475	.25	756	942		166	533	.43	467	606	15
46	298	490		736	931		216	559	.42	441	599	14
47	320	505	.23	719	920		267	584	.43	416	592	13
48	342	519	.25	700	909		318	610		390	585	12
49	364	534		680	898		368	636	.42	364	578	11
50	.65 386	9.81 549	.23	.75 661	9.87 887	.17	.86 419	9.93 661	.43	0.06 339	1.1571	10
51	408	563	.25	642	877	.18	470	687	.42	313	565	9
52	430	578	.23	623	866		521	712	.43	288	558	8
53	452	592	.25	604	855		572	738	.42	262	551	7
54	474	607		585	844		623	763	.43	237	544	6
55	496	622	.23	566	833		674	789	.42	211	538	5
56	518	636	.25	547	822		725	814	.43	186	531	4
57	540	651	.23	528	811		776	840	.42	160	524	3
58	562	665	.25	509	800		827	865	.43	135	517	2
59	584	680	.23	490	789		878	891	.42	109	510	1
60	.65 606	9.81 694		.75 471	9.87 778		.86 929	9.93 916		0.06 084	1.1504	0

130°	Nat.	Log.	Dif.	Nat.	Log.	Dif.	Nat.	Log.	Dif.	Log.	Nat.	49°
		Cosines.			Sines.			Cotangents.			Tangents.	

41°	SINES			COSINES			TANGENTS			COTANGENTS		138°
	Nat.	Log.	Dif.	Nat.	Log.	Dif.	Nat.	Log.	Dif.	Log.	Nat.	
0'	.65 606	9.81 694	.25	.75 471	9.87 778	.18	.86 929	9.93 916	.43	0.06 084	1.1504	60'
1	628	709	.23	452	767		980	942	.42	058	497	59
2	650	723	.25	433	756		.87 031	967	.43	033	490	58
3	672	738	.23	414	745		082	993	.42	007	483	57
4	694	752	.25	395	734		133	9.94 018	.43	0.05 982	477	56
5	716	767	.23	375	723		184	044	.42	956	470	55
6	738	781	.25	356	712		236	069	.43	931	463	54
7	759	796	.23	337	701		287	095	.42	905	456	53
8	781	810	.25	318	690		338	120	.43	880	450	52
9	803	825	.23	299	679		389	146	.42	854	443	51
10	.65 825	9.81 839	.25	.75 280	9.87 668	.18	.87 441	9.94 171	.43	0.05 829	1.1436	50
11	847	854	.23	261	657		492	197	.42	803	430	49
12	869	868		241	646		543	222	.43	778	423	48
13	891	882	.25	222	635		595	248	.42	752	416	47
14	913	897	.23	203	624		646	273	.43	727	410	46
15	935	911	.25	184	613	.20	698	299	.42	701	403	45
16	956	926	.23	165	601	.18	749	324	.43	676	396	44
17	978	940	.25	146	590		801	350	.42	650	389	43
18	.66 000	955	.23	126	579		852	375	.43	625	383	42
19	022	969		107	568		904	401	.42	599	376	41
20	.66 044	9.81 983	.25	.75 068	9.87 557	.18	.87 955	9.94 426	.43	0.05 574	1.1369	40
21	066	998	.23	049	546		.88 007	452	.42	548	363	39
22	088	9.82 012		030	535		059	477	.43	523	356	38
23	109	026	.25	030	524		110	503	.42	497	349	37
24	131	041	.23	011	513	.20	162	528	.43	472	343	36
25	153	055		.74 999	501	.18	214	554	.42	446	336	35
26	175	069	.25	978	490		265	579		421	329	34
27	197	084	.23	958	479		317	604	.43	396	323	33
28	218	098		984	468		369	630	.42	370	316	32
29	240	112		915	457		421	655	.43	345	310	31
30	.66 262	9.82 126	.25	.74 896	9.87 446	.20	.88 478	9.94 681	.42	0.05 319	1.1308	30
31	284	141	.23	876	434	.18	524	706	.43	294	296	29
32	306	155		857	423		576	732	.42	268	290	28
33	327	169	.25	838	412		628	757	.43	243	283	27
34	349	184	.23	818	401		680	783	.42	217	276	26
35	371	198		799	390	.20	732	808		192	270	25
36	393	212		780	378	.18	784	834	.42	166	263	24
37	414	226		760	367		836	859		141	257	23
38	436	240	.25	741	356		888	884	.43	116	250	22
39	458	255	.23	722	345		940	910	.43	090	243	21
40	.66 480	9.82 269	.23	.74 703	9.87 334	.20	.88 992	9.94 935	.43	0.05 065	1.1237	20
41	501	283		683	322	.18	.89 045	961	.42	039	230	19
42	523	297		664	311		097	986	.43	014	224	18
43	545	311	.25	644	300	.20	149	9.95 012	.42	0.04 988	217	17
44	566	326	.23	625	288	.18	201	037		963	211	16
45	588	340		606	277		253	062	.43	938	204	15
46	610	354		586	266		306	088	.42	912	197	14
47	632	368		567	255	.20	358	113	.43	887	191	13
48	653	382		548	243	.18	410	139	.42	861	184	12
49	675	396		528	232		463	164	.43	836	178	11
50	.66 697	9.82 410	.23	.74 509	9.87 221	.20	.89 515	9.95 190	.42	0.04 810	1.1171	10
51	718	424	.25	489	209	.18	567	215		785	165	9
52	740	439	.23	470	198		620	240	.43	760	158	8
53	762	453		451	187	.20	672	266	.42	734	152	7
54	783	467		431	175	.18	725	291	.43	709	145	6
55	805	481		412	164		777	317	.42	683	139	5
56	827	495		392	153	.20	830	342	.43	658	132	4
57	848	509		373	141	.18	883	368	.42	632	126	3
58	870	523		353	130		935	393		607	119	2
59	891	537		334	119	.20	988	418	.43	582	113	1
60	.66 913	9.82 551		.74 314	9.87 107		.90 040	9.95 444		0.04 556	1.1106	0
131°	Nat.	Log.	Dif.	Nat.	Log.	Dif.	Nat.	Log.	Dif.	Log.	Nat.	48°
	COSINES			SINES			COTANGENTS			TANGENTS		

42°	SINES.			COSINES.			TANGENTS.			COTANGENTS.		137°
	Nat.	Log.	Dif.	Nat.	Log.	Dif.	Nat.	Log.	Dif.	Log.	Nat.	
0′	.66 913	9.82 551	.23	.74 314	9.87 107	.18	.90 040	9.95 444	.42	0.04 556	1.1106	60′
1	935	565		295	096		093	469	.43	531	100	59
2	956	579		276	085	.20	146	495	.42	505	093	58
3	978	593		256	073	.18	199	520		480	087	57
4	999	607		237	062	.20	251	545	.43	455	080	56
5	.67 021	621		• 217	050	.18	304	571	.42	429	074	55
6	043	635		198	039		357	596	.43	404	067	54
7	064	649		179	028	.20	410	622	.42	378	061	53
8	086	663		159	016	.18	463	647		353	054	52
9	107	677		139	005	.20	516	672	.43	328	048	51
10	.67 129	9.82 691	.23	.74 120	9.86 993	.18	.90 569	9.95 698	.42	0.04 302	1.1041	50
11	151	705		100	982	.20	621	723		277	035	49
12	172	719		080	970	.18	674	748	.43	252	028	48
13	194	733		061	959	.20	727	774	.42	226	022	47
14	215	747		041	947	.18	781	799	.43	201	016	46
15	237	761		022	936	.20	834	825	.42	175	009	45
16	258	775	.22	002	924	.18	687	850		150	003	44
17	280	788	.23	.73 983	913		940	875	.43	125	1.0996	43
18	301	802		963	902	.20	993	901	.42	099	990	42
19	323	816		944	890	.18	.91 046	926	.43	074	983	41
20	.67 344	9.82 830	.23	.73 924	9.86 879	.20	.91 099	9.95 952	.42	0.04 048	1.0977	40
21	366	844		904	867		153	977		023	971	39
22	387	858		885	855	.18	206	9.96 002	.43	0.03 998	964	38
23	409	872	.22	865	844	.20	259	028	.42	972	958	37
24	430	885	.23	846	832	.18	313	053		947	951	36
25	452	899		826	821	.20	366	078	.43	922	945	35
26	473	913		806	809	.18	419	104	.42	896	939	34
27	495	927		787	798	.20	473	129	.43	871	932	33
28	516	941		767	786	.18	526	155	.42	845	926	32
29	538	955	.22	747	775	.20	580	180		820	919	31
30	.67 559	9.82 968	.23	.73 728	9.86 763	.18	.91 633	9.96 205	.43	0.03 795	1.0913	30
31	580	982		708	752	.20	687	231	.42	769	907	29
32	602	996		689	740		740	256		744	900	28
33	623	9.83 010	.22	669	728	.18	794	281	.43	719	894	27
34	645	023	.23	649	717	.20	847	307	.42	693	888	26
35	666	037		629	705	.18	901	332		668	881	25
36	688	051		610	694	.20	955	357	.43	643	875	24
37	709	065	.22	590	682		.92 008	383	.42	617	869	23
38	730	078	.23	570	670	.18	062	408		592	862	22
39	752	092		551	659	.20	116	433	.43	567	856	21
40	.67 773	9.83 106	.23	.73 531	9.86 647	.20	.92 170	9.96 459	.42	0.03 541	1.0850	20
41	795	120	.22	511	635	.18	224	484	.43	516	843	19
42	816	133	.23	491	624	.20	277	510	.42	490	837	18
43	837	147		472	612		331	535		465	831	17
44	859	161	.22	452	600	.18	385	560	.43	440	824	16
45	880	174	.23	432	589	.20	439	586	.42	414	818	15
46	901	188		413	577		493	611		389	812	14
47	923	202	.22	393	565	.18	547	636	.43	364	805	13
48	944	215	.23	373	554	.20	601	662	.42	338	799	12
49	965	229	.22	353	542		655	687		313	793	11
50	.67 987	9.83 242	.23	.73 333	9.86 530	.20	.92 709	9.96 712	.43	0.03 288	1.0786	10
51	.68 008	256		314	518	.18	763	738	.42	262	780	9
52	029	270	.22	294	507	.20	817	763		237	774	8
53	051	283	.23	274	495		872	788	.43	212	768	7
54	072	297	.22	254	483	.18	926	814	.42	186	761	6
55	093	310	.23	234	472	.20	980	839		161	1.0755	5
56	115	324		215	460		.93 034	864	.43	136	749	4
57	136	338	.22	195	448		088	890	.42	110	742	3
58	157 ·	351	.23	175	436	.18	143	915		085	736	2
59	179	365	.22	155	425	.20	197	940	.43	060	730	1
60	.68 200	9.83 378		.73 135	9.86 413		.93 252	9.96 966		0.03 034	1.0724	0

132°	Nat.	Log.	Dif.	Nat.	Log.	Dif.	Nat.	Log.	Dif.	Log.	Nat.	47°
		COSINES.			SINES.			COTANGENTS.		TANGENTS.		

43°	Sines. Nat.	Log.	Dif.	Cosines. Nat.	Log.	Dif.	Tangents. Nat.	Log.	Dif.	Cotangents. Log.	Nat.	136°
0'	.68 200	9.83 378	.23	.73 135	9.86 413	.20	.93 252	9.96 966	.42	0.03 034	1.0724	60'
1	221	392	.22	116	401		306	991		009	717	59
2	242	405	.23	096	389		360	9.97 016	.43	0.02 984	711	58
3	264	419	.22	076	377	.16	415	042	.42	958	705	57
4	285	432	.23	056	366	.20	469	067		933	699	56
5	306	446	.22	036	354		524	092	.43	908	692	55
6	327	459	.23	016	342		578	118	.42	882	686	54
7	349	473	.22	.72 996	330		633	143		857	680	53
8	370	486	.23	976	318		688	168		832	674	52
9	391	500	.22	957	306	.18	742	193	.43	807	668	51
10	.68 412	9.83 513	.23	.72 937	9.86 295	.20	.93 797	9.97 219	.42	0.02 781	1.0661	50
11	434	527	.22	917	283		852	244		756	655	49
12	455	540	.23	897	271		906	269	.43	731	649	48
13	476	554	.22	877	259		961	295	.42	705	643	47
14	497	567	.23	857	247		.94 016	320		680	637	46
15	518	581	.22	837	235		071	345	.43	655	630	45
16	539	594	.23	817	223		125	371	.42	629	624	44
17	561	608	.22	797	211	.16	180	396		604	618	43
18	582	621		777	200	.20	235	421	.43	579	612	42
19	603	634	.23	757	188		290	447	.42	553	606	41
20	.68 624	9.83 648	.22	.72 737	9.86 176	.20	.94 345	9.97 472	.42	0.02 528	1.0599	40
21	645	661		717	164		400	497	.43	503	593	39
22	666	674	.23	697	152		455	523	.42	477	587	38
23	688	688	.22	677	140		510	548		452	581	37
24	709	701	.23	657	128		565	573		427	575	36
25	730	715	.22	637	116		620	598	.43	402	569	35
26	751	728		617	104		676	624	.42	376	562	34
27	772	741	.23	597	092		731	649		351	556	33
28	793	755	.22	577	080		786	674	.43	326	550	32
29	814	768		557	068		841	700	.42	300	544	31
30	.68 835	9.83 781	.23	.72 537	9.86 056	.20	.94 896	9.97 725	.42	0.02 275	1.0538	30
31	857	795	.22	517	044		952	750	.43	250	532	29
32	878	808		497	032		.95 007	776	.42	224	526	28
33	899	821		477	020		062	801		199	519	27
34	920	834	.23	457	008		119	826		174	513	26
35	941	848	.22	437	9.85 996		173	851	.43	149	507	25
36	962	861		417	984		229	877	.42	123	501	24
37	983	874		397	972		284	902		098	495	23
38	.69 004	887	.23	377	960		340	927	.43	073	489	22
39	025	901	.22	357	. 948		395	953	.42	047	483	21
40	.69 046	9.83 914	.22	.72 337	9.85 936	.20	.95 451	9.97 978	.42	0.02 022	1.0477	20
41	067	927		317	924		506	9.98 003	.43	0.01 997	470	19
42	088	940	.23	297	912		562	029	.42	971	464	18
43	109	954	.22	277	900		616	054		946	458	17
44	130	967		257	888		673	079		921	452	16
45	151	980		236	876		729	104	.43	896	446	15
46	172	993		216	864	.22	785	130	.42	870	440	14
47	193	9.84 006	.23	196	851	.20	841	155		845	434	13
48	214	020	.22	176	839		897	180	.43	820	428	12
49	235	033		156	827		952	206	.42	794	422	11
50	.69 256	9.84 046	.22	.72 136	9.85 815	.20	.96 008	9.98 231	.42	0.01 769	1.0416	10
51	277	059		116	803		064	256		744	410	9
52	298	072		095	791		120	281	.43	719	404	8
53	319	085		075	779	.22	176	307	.42	693	398	7
54	340	098	.23	055	766	.20	232	332		668	392	6
55	361	112	.22	035	754		288	357	.43	643	385	5
56	382	125		015	742		344	383	.42	617	379	4
57	403	138		.71 995	730		400	408		592	373	3
58	424	151		974	718		457	433		567	367	2
59	445	164		954	706	.22	513	458	.43	542	361	1
60	.69 466	9.84 177		.71 934	9.85 693		.96 569	9.98 484		0.01 516	1.0355	0

133°	Nat.	Log.	Dif.	Nat.	Log.	Dif.	Nat.	Log.	Dif.	Log.	Nat.	46°
	Cosines.			Sines.			Cotangents.			Tangents.		

44°	SINES			COSINES			TANGENTS			COTANGENTS		135°
	Nat.	Log.	Dif.	Nat.	Log.	Dif.	Nat.	Log.	Dif.	Log.	Nat.	
0'	.69 466	9.84 177	.22	.71 934	9.85 693	.20	.96 569	9.98 484	.42	0.01 516	1.0355	60'
1	487	190		914	681		625	509		491	349	59
2	508	203		894	669		681	534	.43	466	343	58
3	529	216		878	657		788	560	.42	440	337	57
4	549	229		853	645	.22	794	585		415	331	56
5	570	242		833	632	.20	650	610		390	325	55
6	591	255	.23	813	620		907	635	.43	365	319	54
7	612	269	.22	792	608		963	661	.42	339	313	53
8	633	282		772	596	.22	.97 020	686		314	307	52
9	654	295		752	583	.20	076	711	.43	289	301	51
10	.69 675	9.84 308	.22	.71 732	9.85 571	.20	.97 133	9.98 737	.42	0.01 263	1.0295	50
11	696	321		711	559		189	762		238	289	49
12	717	334		691	547	.22	246	787		213	283	48
13	737	347		671	534	.20	302	812	.43	188	277	47
14	758	360		650	522		359	838	.42	162	271	46
15	779	373	.20	630	510	.22	416	863		137	265	45
16	800	385	.22	610	497	.20	472	888		112	259	44
17	821	398		590	485		529	913	.43	087	253	43
18	842	411		569	473	.22	586	939	.42	061	247	42
19	862	424		549	460	.20	643	964		036	241	41
20	.69 883	9.84 437	.22	.71 529	9.85 448	.20	.97 700	9.98 989	.43	0.01 011	1.0235	40
21	904	450		508	436	.22	756	9.99 015	.42	0.00 985	230	39
22	925	463		488	423	.20	813	040		960	224	38
23	946	476		468	411		870	065		935	218	37
24	966	·489		447	399	.22	927	090˙	.43	910	212	36
25	987	502		427	386	.20	984	116	.42	884	206	35
26	.70 008	515		407	374	.22	.98 041	141		859	200	34
27	029	528	.20	386	361	.20	098	166		834	194	33
28	049	540	.22	366	349		155	191	.43	809	188	32
29	070	553		345	337	.22	213	217	.42	783	182	31
30	.70 091	9.84 566	.22	.71 325	9.85 324	.20	.98 270	9.99 242	.42	0.00 758	1.0176	30
31	112	579		305	312	.22	327	267	.43	733	170	29
32	132	592		284	299	.20	384	293	.42	707	164	28
33	153	605		264	287	.22	441	318		682	158	27
34	174	618	.20	243	274	.20	499	343		657	152	26
35	195	630	.22	223	262		556	368	.43	632	147	25
36	215	643		203	250	.22	613	394	.42	606	141	24
37	236	656		182	237	.20	671	419		581	135	23
38	257	669		162	225	.22	728	444		556	129	22
39	277	682	.20	141	212	.20	786	469	.43	531	123	21
40	.70 298	9.84 694	.22	.71 121	9.85 200	.22	.98 843	9.99 495	.42	0.00 505	1.0117	20
41	319	707		100	187	.20	901	520		480	111	19
42	339	720		080	175	.22	958	545		455	105	18
43	360	733	.20	059	162	.20	.99 016	570	.43	430	099	17
44	381	745	.22	039	150	.22	073	596	.42	404	094	16
45	.70 401	758		019	137	.20	131	621		379	088	15
46	422	771		.70 998	125	.22	189	646	.43	354	082	14
47	443	784	.20	978	112	.20	247	672	.42	328	076	13
48	463	796	.22	957	100	.22	304	697		303	070	12
49	484	809		937	087		362	722		278	064	11
50	.70 505	9.84 822	.22	.70 916	9.85 074	.20	.99 420	9.99 747	.43	0.00 253	1.0059	10
51	525	835	.20	896	062	.22	478	773	.42	227	052	9
52	546	847	.22	875	049	.20	536	798		202	047	8
53	567	860		855	037	.22	594	823		177	041	7
54	587	873	.20	834	024	.20	652	848	.43	152	035	6
55	608	885	.22	813	012	.22	710	874	.42	126	029	5
56	628	898		793	9.84 999		768	899		101	023	4
57	649	911	.20	772	986	.20	826	924		076	017	3
58	670	923	.22	752	974	.22	864	949	.43	051	012	2
59	690	936		731	961	.20	942	975	.42	025	006	1
60	.70 711	9.84 949		.70 711	9.84 949		1.0000	0.00 000		0.00 000	1.0000	0

134°	Nat.	Log.	Dif.	Nat.	Log.	Dif.	Nat.	Log.	Dif.	Log.	Nat.	45°
		COSINES.			SINES.			COTANGENTS.		TANGENTS.		

www.ingramcontent.com/pod-product-compliance
Lightning Source LLC
Chambersburg PA
CBHW021523090426
42739CB00007B/744